£1.50

Peasant Studies in Nepal

Peasant Studies
in
NEPAL

Editors

Laya Prasad Uprety
Suresh Dhakal
Jagat Basnet

VAJRA
BOOKS Inc.

Published by
Vajra Books Inc.
Sunnyside, New York
USA
e-mail: vajrabooksny@gmail.com

Distributed by
Vajra Books
Jyatha, Thamel, P.O. Box 21779, Kathmandu, Nepal
Tel.: 977-1-4220562, Fax: 977-1-4246536
e-mail: bidur_la@mos.com.np
www.vajrabooks.com.np

This is a peer-reviewed publication.

ISBN 978-1-64316-089-4

Photograph Courtesy (Front and Back cover):
Community Self Reliance Center (CSRC)

Cover Design, Layout and Printed by
Dongol Printers, Kathmandu, Nepal. e-mail: dongolprinters@gmail.com

Foreword

Studies of the peasant societies and cultures have historically been the foci of socio-cultural anthropology at the global, regional, national, sub-national and local levels.Given the fact that there is the panoply of real differences on the 'peasantry' of Nepal across three major geo-ecological belts, developing generalizations on them is an uphill task for social scientists in general and anthropologists in particular. The available literature on 'peasantry' in Nepal fails to describe the diversity within 'peasantry' on the one hand and explicate it in the changing context largely triggered by the neo-liberal model of economic development on the other hand. 'Peasantry' as a distinct term has been in vogue in academia and in everyday communication for several years in Nepal despite its relatively poor definition, description and analysis with empirical substantiations. This is demonstrative of the fact that studies related to 'peasants' and 'peasant economy' have remained relatively less prioritized in the academia (such as in anthropology despite its disciplinary maturity) but concerned government departments and non-governmental agencies claim that they work for the betterment of the peasants in the country. Contextually, academicians and development professionals interested to study 'peasants' and 'peasant economy' have to be effortful to look for answers of a battery of research questions. More specifically, these would comprise the following questions: who are 'peasants' in our context?; how is 'peasant economy' changing along with the changes in land tenure system and other land policies?; is 'peasant' a distinct class category?; what are the socio-economic and cultural signifiers of 'peasant'?; what has happened to them when market has penetrated into their traditional domestic mode of production?; how has the 'peasantry' been a political factor in the past?; what kind of gender relations can we discern among the 'peasants'; how does it differ (or does it not differ) from the 'indigenous' and 'tribal' groups that practiced customary and communal land tenure system?; why is the 'peasantry' on decline with the rise of neo-liberal economy? or what are the triggers of the 'depeasantization' in contemporary Nepali society?; how are 'peasants' organized for their land rights over the years under the empowerment framework?; how are the agenda of peasants' upliftment hijacked by the political parties?; etc. These and many other questions are yet to be answered with rigorous theoretical orientations and empirical evidences.

Against this backdrop, the Central Department of Anthropology (CDA),Tribhuvan University and Community Self-reliance Centre (CSRC), Kathmandu had organized a thematic seminar collaboratively on "Conceptualizing Peasantry: Understanding Peasants and Peasant Economy in Nepal" on 28 July, 2017. This maiden collaborative effort of an academic institution and a civil society organization (CSO) brought the issues of 'peasants' and 'peasant economy' at the centre stage of social science research, discussion and academic engagement. However, both organizers would appreciate and welcome more organizations and institutions to be a part of this collaborative effort if there are interests and willingness to contribute to this process of 'knowledge building' in the days to come. As expected, social science researchers, experts, intellectuals, advocates and development practitioners had joined this common forum of academic engagement embedded with the objective of enhancing our collective understanding on our society and economy with special reference to the transformations of the 'peasantry' and 'peasant economy'.

The papers presented in the seminar have been published in the form of edited volume of book entitled '**Peasant Studies in Nepal**'. Therefore, we, on behalf of CDA and CSRC, would like to thank all the eight eminent scholars for their academic paper contributions. Similarly, we would thank the editors for their meticulous editing of the book. We are sanguine that the book will be useful to graduate students of the social sciences, researchers, policy makers, government program implementers, development professionals, and activists of peasants' rights.

Jagat Deuja
Executive Director, Community Self-Reliance Centre

Prof. Laya Prasad Uprety, Ph.D
Head, Central Department of Anthropology
Tribhuvan University, Kathmandu

May, 2018

Editorial Preface

The book is an academic attempt to fill the gap of literature on the contemporary peasant society of Nepal which is undergoing rapid transformations due to the penetration of global capitalism—a function of the adoption of neoliberal economic development model by Nepal after the democratic resurgence in 1990 A.D under the dictates of multilateral development agencies of which Nepal has been a member for long. Equally important is the genesis of an institutional effort between an academic institution and a civil society organization for the "production of knowledge" on 'peasants', 'peasant economy' and 'peasants' rights' in the context of Nepal and South Asia. This book has been the outcome of the contributions made by five anthropologists, two sociologists and one agricultural scientist. The book contains eight chapters on peasant studies.

Dr. Suresh Dhakal, in his chapter entitled "**Peasant as Anthropological Category and Ethnographic Subject: A Critical Overview of Peasantry and Peasant Politics**", analytically dwells on peasants whose lives are governed by the agrarian themes and are defined differently in different spatio-temporal contexts. Interestingly, he is in agreement with A.L Kroberʼs notional definition of peasants' societies as "part-societies with part-cultures". He argues that 'peasants' are often understood someone as 'farmers'; but more specifically, 'peasants' have a social position, bound to their land. More specifically, the term "peasant" also describes a lifestyle and social status, whereas the term "farmer" would refer more to one's occupation, a profession but not a social position. However, it is largely contextual and partly contentious too. Therefore, a context is needed to define and describe what a 'peasant' is. Based on reviews of relevant literature (both general and anthropological literature in particular), he presents an overview of peasants and peasantry as various authors have dealt with them in the past. The author also tries to unravel the intricate relationship between peasantry and politics, one of the fundamental issues of discourses on the peasant question in history. Literature on peasantry, and peasants and their political connections have demonstrated 'how peasants come to take part in political mobilization and how they are affected by such participation'. While considering the peasantry as a political f actor, the author focuses more on peasants' resistance, and political mobilization of peasants by the political parties. Finally, he discusses on how to approach peasants' movement in general and presents an overview of peasants' resistance and political mobilization in Nepal in particular. Peasants' movements are often associated with exploitation and disobedience frameworks; however, he agrees with the cautionary note of J. Scott (1976) who asserted, "Growing exploitation of the peasantry may well be a necessary cause of rebellion, but it is far from a sufficient cause". Hence, he has endeavoured to unpack some of the intricate, overt or concealed link between peasants' uprising and political development and politicization of peasantry in Nepal. This is also because Nepal's peasants' movements also experienced the similar relationship with the politics and burgeoned along with the political process and movements of the left politics in particular. The focus on peasant organization and urge for struggle was a long sustained feature of communist parties in the history of Nepal since 2006 . Thus, his review on Nepal's peasant movements and their political linkages correspond to Wolford (2007), when he says, "Micro-politics articulates with macro-politics in the sense that the strength of the individual movements draws on the ability of members to recognize and connect their particular conditions and political projects". Hence, this review paper is an effort to examine peasant as 'anthropological category' and 'ethnographic subject' in order to develop the analysis as the 'anthropology of peasantry' and 'peasant insurgency' in relation to its politicization.

Dr. David Seddon, in his chapter entitled "**Resistance, Rebellion, Revolt and Revolution: A Historical/Anthropological Consideration of Peasant Movements and Other Forms of Rural Unrest in Nepal**", reminds the readers of the introduction to 'The People's War in Historical Context' in which he and his fellow colleague from Nepali scholarship had commented that 'there is a long but largely undocumented history of popular uprisings and revolts against the autocratic rulers of Nepal... '(Karki & Seddon, 2003, p.3). Since then, he argues that there have been several contributions to such a history, to which he refers in this chapter, but in this first part of a new consideration of the history of rural unrest in Nepal, he attempts to bring together what is known regarding the period from the 'unification' of the Gorkhali State to the beginning of the People's War.

Dr. Janak Rai, in his chapter entitled "**The Landlord State, *Adivasi* People, and the Escape Agriculture in the Eastern *Tarai* of Nepal: A Historical Analysis of the Transformation of *Dhimals* into a Farming Community**", has used the concepts of '*adivasi*' and 'peasant' as analytical categories to emphasize distinct societies with respect to their relationships to the land, nature of economic system, and relationships with the state over the control of land. He has argued that if we consider '*adivasi* and 'peasant' as analytical categories, then the centrality of 'land' in defining the two categories become obvious. He has further cogently argued that 'land' is an important material possession (a physical entity, property, a resource) with its cultural and symbolic relevance for peasants. For the '*adivasi*', land is a defining feature of their collective identity but it is primarily an inalienable wealth, not a commodity. Finally, his analysis shows that the peasant-land relationship draws heavily on capitalist ontology of commodity and property while the *adivasi*-land relationships underscore the total embeddedness and mutual production of land, people and culture in totality.

Mr. Jagat Basnet, in his chapter entitled "**Marginalization of the Tenants in Nepal: A Political-Economic Analysis**", proffers a trenchant critique on the marginalization of tenants in Nepal despite the promulgation and implementation of the much-trumpted land reform measure under the Lands Act of 1964. He has argued that tenants are the most "marginalized class" for generations in Nepal. They are directly dependent on landlords or richer class for their subsistence. The rulers, in the past, distributed the state land to their henchmen, supporters and relatives and Hindu ecclesiastics who were not the peasants which triggered the origin of the "tenancy system" and "absentee landlordism". Despite the six amendments in the 1964 Lands Act, more than five hundred thousand tenants have been deprived of their tenancy rights and similar number of tenants has already been evicted. The 1964 land reform programme was introduced and implemented in three phases embedded with the objective of distributing of the excess land above ceiling to landless and tenant peasants. In reality, only one percent of the total expropriated land was distributed to landless and tenant farmers. This largely benefited only richer people and staff of land reform offices. The ownership of land and its control by marginalized farmers has been a far-fetched dream and their economic condition has not changed even after nearly five decades of the implementation of land reform initiative. 'Land' and 'labour' have been recognized as commodities and thus marginalized tenants have to depend on landlords. Historically, landless and tenant farmers have been providing surplus to the ruling and landed class but they have been buried in debt and hence, have low economic status. Although the political system has changed, there has only been little change in the lives of poor and tenants. Instead, new forms of exploitation have emerged. There is one discernible change, that is, poor people used to work as tenants previously and now they are working as "sharecroppers" or "wage labourers". In the past, the cultivating class did not have access to education, and hence, they remained largely non-literate and uneducated. They were compelled to provide *corvee* to landlords and the state, and pay exorbitant interest rates for loans, high rates of agricultural rents (of the tilled land) and different kinds of levies to landlords or village chiefs. These feudatory obligations caused tenants and sharecroppers to become victims of a vicious cycle of poverty. The principal objective of this study is to analyze and explicate how macro and micro level policy practices and political systems have historically triggered the marginalization of tenants. More specifically, he focuses on the critical analysis of the marginalized situation of tenant peasants and sharecroppers in Nepal.

Mrs. Yamuna Ghale, in her chapter entitled "**Peasantry and State-People Relations in Food Security Governance: Exploring the Linkages from Gender Perspective**", has considered farmers in general and women farmers in particular as an "operational category of peasantry". She has analyzed the interrelationship between state as a "duty bearer" and people (male and female farmers) as "rights holders" focusing on accountability and participation as principles of the food security governance. While examining this relation, she has focused on accountability provisions stipulated in the Constitution of Nepal (2015), Food Act (1992), and Consumer Protection Act (1998) which are some of the major framework documents to ensure Right to Food (RtF) and Food Security Governance (FSG). In addition, for examining people's engagement, she has analyzed provisions of participation of peasants, especially women's involvement in specific legal and policy documents such as the Constitution of Nepal (2015) and Agriculture Development Strategy (ADS) which is being implemented. She has identified major gaps and opportunities to promote gender sensitive accountability and participation in food security governance. In doing so, she has used the information and data collected for her ongoing Ph.D research on 'Food Security Governance and Right to Food: Understanding State-People Relationship in Nepal'. She argues that governance has great significance in ensuring inclusion of peasants for broad economic growth and agrarian transformation in the country. It is important to ensure progressive realization of food for all, in which the state has to create enabling environment by meaningfully supporting small-holder farmers; harmonizing food safety standards; enabling and regulating market space; reducing environmental impacts; balancing production and trade as well as facilitating emergency food supplies. She concludes that ensuring food security and right to food requires meaningful participation of people with clear provision for mobilization of private sector, civil society organizations, and family as an institution to take up their function as responsibility bearers. The rights and entitlements can only be enjoyed when the accountability mechanisms are explicitly defined, sincerely implemented and regularly monitored by the State.

Dr. Madhu Giri, in his chapter entitled "**Changing Labour Regimes of the *Musahar* Peasants: A Study from the *Tarai* of Nepal**", presents the analysis on the transition of agricultural labour regimes of the *Musahar* community from *Haruwa-Charuwa* (a system of unfree agricultural labourers) to sharecropper and contractual market labourers. He has explored the changing form of (de)peasantry as an elementary form of labour heterogeneity from "full-time regular work" to "part-time irregular contract". Until the recent decades, most of the adult *Musahars* had eked out their living by being *Haruwa-Charuwa* and they were well known as the "best farm labourers" in the central-eastern *Tarai*. He has shared his observation that neither *Haruwa-Charuwa* system nor the big landlordism was the dominant form of production; rather intermediate landholding and sharecropping, and/or heterogeneous contractual labour systems were the dominant forms of peasantry in the study areas. Sharecropping has been popularized as an "equal relationship" whereby the cultivator and landlord provide equal inputs, take equal risks and receive equal share of the crops produced. The most common division mentioned by sharecroppers and other sources is *Adhiyar*, that is, half the crops produced goes to the sharecropper and half to the landlord. Different forms of sharecropping were closely related to sharing of inputs, and type of land, livestock which were, in turn, influenced by the circumstances under which the sharecropping developed. The landless *Musahars* retained various forms of sharecropping not only to utilize them for their households and reserved labourers but also to keep hold of social relations. The most common talking subject is that there is a dearth of agricultural labourers in central-eastern *Tarai* region. Contractual labour and labour migration among all caste/ethnic groups and lower prestige of farm labourers triggered a dearth of farm labourers in the district. The *Musahars* also enjoyed off-farm wage labour which was organized through bargaining and contract. Land and farm embedded social relations were at the verge of crisis. Because of changes in politico-economic labour policies and application of uncertain heterogeneous off-farm livelihood strategies, they were dragged in the world of semi-proletariat or what Anna L. Tsing (2015) termed pericapitalist freedom in which freed community has no long-term economic plan. This article argues that when agricultural labour became less important in the livelihood portfolios, processes of a particular social reproduction or production ruptured traditional boundaries and came into new terrain of transition and resultant economic precariousness.

Dr. Kapil Babu Dahal, in his chapter entitled "**Discourses on the Transformation of the Peasantry: Looking through the Life Histories from Dullav Area**", has asserted that subsistence agriculture in the middle hills of Nepal has historically reflected a basic feature of peasantry. Considering its perpetual reliance on simple agrarian technology and mostly of subsistence nature of farming, often, peasantry in Nepal has been represented as a stagnant social milieu. Nevertheless, its in-egalitarian feature has been acknowledged by the political activists and scholars working in the areas of peasantry. He has dwelled on how societal transformations are taking place among peasants and their ways of life in the central hills of Nepal over the last three decades. More specifically, his analysis revolves around one fundamental question, that is, "are such changes limited to 'the differentiation of peasantry' leading to the proliferation of capitalist agricultural development as analyzed by V.I. Lenin or do they have some local specificities in Nepali context? Comprehending the consequences of migration, both at the domestic and international levels, his analysis has demonstrated that transformation in peasantry in Nepal has its own specificities. For the substantiation of his argument, he has dealt with mainly on the prominent three areas of social life of the peasants from the study site, viz. their changing relation with land, their experiences and consequences of migration and the emerging consumption pattern.

Dr. Laya Prasad Uprety, in his chapter entitled "**Peasants' Land Rights: A Perspective from Engaged Anthropology**", argues that in a social world characterized by "inequalities" and "injustice", anthropologists cannot be "value-neutral" and maintain the "detachment" from the cultural and structural analysis as per the positivistic cannons of research. Following Nancy Scheper-Hughes (1995) who proposes a "militant anthropology", we can no longer be content with the classical "cultural relativistic" analysis because we cannot keep our eyes closed on the "social injustice", "inequalities", and "gross violation of human rights" in the given socio-cultural system under study. Hence, we should continue to be concerned when basic human rights are denied to the people. Granted this ideological position, anthropologists need to engage themselves as the "social critics", "anthropological citizens", "advocates", "collaborative ethnographers", "activist researchers", "active agents of change", "decolonizers" of the relationships between researchers and informants, "human liberators", etc. In a nutshell, anthropological study cannot be "apolitical". He has made a modicum of efforts to analyze peasants' "land rights" as "human rights" at the international and South Asian regional context by subsuming the analysis of the land rights movement cases of Nepal, India and Bangladesh.

Globally accepted concepts of human rights (civil, political, economic, social and cultural) enshrined in the declaration and covenant of the United Nations General Assembly have always been the concern of millions of landless and land-poor peasants the world over – a function of the failure of their progressive realization by the wretched people even in a protracted period of 70 years or so. Notwithstanding the fact that international and national efforts made after 1948 have, to a large measure, played an instrumental role for creating the grounds to help people enjoy their human rights, much remains to be done to ensure social equity and justice in the world where people controlling vast lands and their natural resources by "foul means" within the national boundaries are the ones controlling the national governments and international economic organizations where they monopolize the political decision-making vis-à-vis the control and allocation of these resources. The resultant resource inequity functions as a bottleneck triggering the political, economic, social and cultural disempowerment of the vast number of peasants (landless, tenants, marginalized women farmers, and indigenous people) whose 'culture and survival' are inextricably linked to land and its natural resources. Given this context, mainstreaming land rights as human rights is an urgent concern in the South Asian context because land rights, as elsewhere, is a gateway to the progressive realization of other human rights (given the fact that land resource inequity has been continuing in the region despite the institutional efforts have been made in the past as mere 'lip service' by the bragadocious politicians of the region to address it).

Given the fact that the social charter of South Asian Association for Regional Cooperation (SAARC) has the regard for civil society organizations (CSOs) by reaffirming the need to develop a regional dimension of co-operation in the social sector and has asserted principles that 'members of the civil society uphold for equity and social justice; respect for and protection of fundamental rights, respect for diverse cultures and people-centred development', land rights issues can also be brought in the People's SAARC (South Asian People's Summit, People's Assembly), South Asia Forum and Working Group on South Asia Human Rights Mechanism (the only existing mechanism) for debates and discussions apropos of the land issues and its findings can be communicated to the SAARC governments through the SAARC secretariat located in Kathmandu, Nepal despite the fact that SAARC seems to be in the nascent stage for the enforcement of human rights (including the issues of land rights). The regional network of CSOs has to play a proactive role to engage the Technical Committee on Agriculture and Rural Development (TCARD) of the SAARC created in 2006 (to address challenges of ensuring the food security and nutritional security at the level of agricultural ministers). What is of paramount importance at this stage is the creation of an environment in which the government representatives of all member countries of the SAARC can participate in the dialogues of land rights issues raised by the prominent CSOs and representatives of victims of the gross violation of land rights in the region. The goal of this task must be to convince the SAARC authorities to create permanent inter-governmental human rights mechanisms such as the Regional

Committee on the Issues of Land Rights and Special Rapporteur on Agrarian Issues and Land Rights of Indigenous Peoples. The success of the creation of such mechanisms may contribute to lessening the gross violation of land rights in the region and ensuring the enjoyment of civic, political, economic, social and cultural rights of the people. Enjoyments of these rights by the poor peasants will pave the path for realizing the Sustainable Development Goals (SDGs) germane to "poverty", "hunger", "gender equality" and "life on land" by 2030.

Laya Prasad Uprety
Suresh Dhakal
Jagat Basnet

May, 2018

Table of Contents

Notes on Contributors

DR. SURESH DHAKAL is the Lecturer at the Central Department of Anthropology, Tribhuvan University, Kathmandu, Nepal. He specializes in political anthropology. He has contributed to agrarian and peasant studies with the publication of a number of research articles to his credit. Of late, his research interest is also on 'poverty dynamics'.

DR. DAVID SEDDON is the retired Professor of Politics and Sociology at the School of Development Studies, University of East Anglia, the UK. He specializes in political economy.He has co-authored two major classic books with the adotion of political-economic approach, namely, "Nepal in Crisis", and "Peasants and Workers in Nepal". Of late, he has also researched about the Muslims of Nepal and published a book on them.

DR. JANAK RAI is the Associate Professor at the the Central Department of Anthropology, Tribhuvan University, Kathmandu, Nepal. He has specialized in the *Dhimals* of eastern *Tarai* of Nepal. He has contributed to the studies of ethnicity and identity. He has also his academic interest in historical anthropology.

MR. JAGAT BASNET is the doctoral candidate in Sociology at the Rhodes University, South Africa. He is also the Research and Policy Adviser to the Community Self-Reliance Centre, Kathmandu, Nepal. He has conducted empirical studies on peasant and agrarian studies and has been extensively involved in the movements of land and agrarian issues and policy advocacy.

Mrs.Yamuna Ghale is the doctoral candidate at the Faculty of Agriculture of the Agriculture and Forestry University, Nepal. She has also been associated with Swiss Development Corporation (SDC) in Kathmandu for long in the capacity of a senior development professional. Her research is centered on "food security" and "feminization of agriculture".

Dr. Madhu Giri is the Lecturer at the Central Department of Anthropology,Tribhuvan University, Kathmandu, Nepal. He specializes in the *Musahars* of the *Tarai* of Nepal with political economic perspective within the sub-field of political anthropology and has a number of research articles published on them to his credit.

Dr. Kapil Babu Dahal is the Lecturer at the Central Department of Anthropology, Tribhuvan University, Kathmandu, Nepal. He specializes in medical anthropology and has published a number of research articles in this subfield to his credit.

Dr.Laya Prasad Uprety is the Professor and Head of the Central Department of Anthropology, Tribhuvan University, Kathmandu, Nepal. He specializes in the anthropology of natural resource management with focus on common property regime in water and forest resources.He has contributed substantially to the studies of farmer-managed irrigation system and land and agrarian rights movement in Nepal.

Peasant as Anthropological Category and Ethnographic Subject: A Critical Overview of Peasantry and Peasant Politics

SURESH DHAKAL

Peasant and Anthropological Concerns

In anthropology, the study of peasants as a distinct anthropological category, 'its inception, growth, and demise' as someone would argue is 'inseparable from an evaluation of the broader intellectual context of social anthropology in which it is found' (Kearney, 1996, p.1). This paper tries to unravel some of those historical contexts and complexities in studying peasants, and peasant politics. Contextually, it is difficult to generalize about peasantries given the very many real differences among them and the absence of distinctive economic criteria making off peasants from tribal" (Dalton, 1972, p.387). Sutti Ortiz characterised peasants as, "traditionally-oriented and slow to change their patterns of behaviour" (Ortiz, 1987, p.300) who were essentially the "rural agriculturalists of extremely low socio-economic and political status" (Powel,1971, p.1). Nevertheless, it has been consistently maintained that, "peasant culture is complex and many-sided and that peasant economy is unique" (Watters, 1994, p.24).

Generally, 'subsistence cultivators' (Nurge,1971, p.1260) whose 'surpluses [taxes] are transferred to dominant group of rulers' (Wolf,1966, p.3) and thus have remained 'diagnostic core' of peasatry. Shanin (1990) carried it further as "the use of land is a necessary and generally sufficient condition to enter the occupation, and acts, therefore, as an entrance ticket to the peasantry" (Shanin, 1990, p.24). And "[t]he family farm is the basic unit of peasant ownership, production, consumption and social life" (Shanin, 1990, p.25). Following Dalton (1972) and Waters (1994), Johnson et al. (2005) maintain "...there is no such thing as a generic peasantry, but rather peasants exist in a multiplicity of forms and circumstances, the product of local settings and shaped by their histories"(Johnson et al., 2005, p.946).

One of the earliest, more comprehensive but encompassing definition of peasants by Wolf (1969) portrays 'as strictly as possible' that peasants are 'population that are existentially involved in making autonomous decisions about cultivation' (as quoted in Kearney, 1996, p.2). Earlier, Wolf (1955) had formulated three basic criteria of peasants as a social type with a view of 'Latin American condition'. For him, the first criterion remains, 'primary involvement in agricultural production, where he draws line between peasants, on the one hand, and fishermen, strip miners, rubber growers, and livestock keepers, on the other (Wolf, 1955, p.453) whereas Firth (1952) had defined the term as widely as possible, including not only agriculturalists but also fishermen and rural craftsmen (1952, p.87). Likewise, Wolf's second criterion was 'effective control of land', where he argued to exclude 'the tenants whose control of land is subject to an outside authority' and his third criterion was 'a primary orientation toward subsistence, including fulfilling the needs defined by culture rather than reinvestment' (Wolf, 1955, pp.453-454). Hence, he argued that the term peasants a 'structural relationship, not a particular culture content' (Wolf, 1955, p.454).For example, Nurge puts it as "[A] peasant is a peasant does" (Nurge, 1971, p.1260). Kearney (1996) takes the shift overtime from "anthropology of peasants as unitary objects to peasants as complex subjects which marks a maturation of peasant studies" (Kearney, 1996, p.7).

Adding to definitional problem of peasants, Johnson (2005) argues, "what Marx (1969, pp.123-128) had to say about peasants in France in the first half of the nineteenth century, those famous and much quoted passages at the end of *The 18th Brumaire of Louis Bonaparte*, cannot be directly extrapolated to characterize all peasantries at all times and places" (Johnson et al., 2005, p.946) where Marx comprehended peasantry as a class (see Marx [1950] 1987,pp. 331-337).

Redfield (1960), locating peasantry in an ecological context and social structure, defines a peasant as "...a man who is in effective control of a piece of land...This does not require of the peasant that he owns the land or that he has any particular form of tenure or any particular form of institutional relationship to the gentry or the townsmen" (Redfield, 1960, p.19). However, a need for a systematic understanding of the complexities of the peasantry" (as quoted in Dalton, 1972, p.385) was yet to be done. Dalton's 'Peasantries in Anthropology and History' was the response and the efforts towards that direction where he suggests the need for a combination of historical, economic and anthropological analysis (Dalton, 1972). He summarzes peasants as:

> "...low-class farmers in a sharply stratified system...low political and social status...tied to dependent land tenure within a larger traditional system in which peasants...had no real alternative means of livelihood. Daily subsistence and emergency support depended on their maintenance of their client position with socioeconomic superiors...peasants of all time and places are structured inferior" (Dalton, 1972, p. 406).

Given the complexities and extensive diversities between the peasantries, Shanin (1987) puts in a more detailed and elaborate form where he distinguishes peasantry as:

> "[S]mall agricultural producers, who, with the help of simple equipment and the labour of their families, produce mostly for their own consumption, direct or indirect, and for the fulfilment of the obligations to holders of political and economic power"(Shanin, 1987, p.3; 1990, p.23).

Hence, Shanin further identifies peasantry with their "...specific relation to land, the peasant family farm and the peasant village community as the basic unit of social interaction, a specific occupational structure and particular patterns of past history and contemporary development". This leads further to some "particularities of peasants' position in society and political action typical of it" (Shanin, 1990, p.24).

Hence, there may be different definitions but nearly all authors agree that "peasant societies are..." to use Kroeber's words, "part-societies with part-cultures" (Kroeber, 1948, p.284; Keyes,1866, p.2). But Kearney (1996) is 'intentionally critical' towards 'anthropological essentialization of the peasant' as he argues 'the peasant opens new horizon for modern anthropology but as a type is powerfully disruptive of its classic structure threatening its epistemological political bases" (Kearney,1996, pp.5-6). Watters (1994) reviewing the corpus of literature on peasantry produced by predecessors categorised five different conceptual approaches to study the 'peasantry'[1], and concludes that each provides some 'illuminations in the search of understanding and explanation, but at the same time, each tends to be myopic' (Watters,1994, p.24). Hence, defining and approaching peasantry is always a contested and contextual issue among the anthropologists.

I do not engage much in debate in defining 'peasants'. However, to avoid misleading simplification, we follow Galjart's definition as, "all those more or less poor people who earn their living in agriculture, including those who work on the land solely for payment of money... small holders and agricultural labourers"(Galjart, 1976, p.3). Hence, we use the term 'peasant' simply to denote 'a person who lives in the rural area and works on the land, especially as a smallholder, tenant cultivator or an agricultural labourer' by taking Firth's (1951, p.87) identification into consideration. In economic terms, a 'peasant' is presumably a man

---

[1] He devotes an entire chapter on Approaches to Peasantry (page11-25) and summarized them into five categories: (i) the ethnographic cultural tradition; (ii) the Durkheimian tradition often allied to functionalist sociology; (iii) the 'specific economy' approach; (iv) the Marxist tradition of class analysis and the dependency approach, and, (v) the ethno-historical approach (Watters, 1994,p.11).

who produces – usually through cultivation – mainly for his own household's consumptions, but who also produces something to exchange in a market for other goods and services. [Hence], by a peasant economy, one means a system of small-scale producers, with the simple technology and equipment often relying primarily for their subsistence on what they themselves produce. The primary means of livelihood of the peasants is cultivation of the soil.

Kishan is a literal translation that is common in use in Nepali; however, with a caution as it sometimes becomes discursive and floating, as the term signifies both landed class and the tenants and small holders. Therefore, even within Nepal, *Kishan* is not a single homogenous category also because their formation has been shaped by different political-economic, environmental, geographical and historical contexts. Poudyal (2016) even uses a term 'ethnic peasantry' for being the fact that most of the peasants belonged to ethnic category; however, 'ethnic peasantry' (Poudyal 2016) as a separate analytical category does not hold much strength. Therefore, we rely more on Giljart's (1976) and Firth's (1951) definitions as they well fit to our contexts and are more inclusive of all.

However, a critical engagement in defining and typifying the term 'peasant' or '*kishan*' is a prerequisite; hence, more anthropological involvement is needed in this context. As we know, in Nepal's context, "[A]s in the case with the peasants in this country, the study on peasantry has also been marginalized" (see Tuladhar (2046 BS [1991]). Since then, a little progress has been made in this regard which is utterly inadequate. In anthropology, in particular, leaving a few exceptions of some scattered and patchy works, peasantry has remained one of the least studied areas by natives or foreign scholars focussing specifically on peasant issue in Nepal; let alone the studies on peasant movements, the revolt against the landlords and their protector, the state. A few examples are there where anthropologists have touched upon the subject while discussing other issues. *Peasants and Workers in Nepal* by Seddon, Blaikie & Cameron (1979) is of course an exemplary one. 'Concept of Economy in a Peasant Society' by Dahal (1981) is yet another brief but a pioneering work on peasant economy in Nepal by a native anthropologist.

Peasants' Resistance, Political Mobilization, and Peasantry as a Political Factor

To concur with Roy (2006), "One of the fundamental issues of discourse on the peasant question in history has hinged on the pivotal idea of political mobilization of peasantry…" (2006,p.99). For an anthropologist, "…how peasants come to take part in political mobilization and how they are affected by such participation" (Wolf, 1975, p.385) is an interesting question to ask. He observes, "[T]he entry of rural population into the political arena – dramatized by the Chinese and Cuban revolutions, and by the war in Vietnam – also raised questions about the political role of peasantry, about peasant politics, peasant ties with the state, peasant leadership, peasant readiness and reluctance to engage in rebellion, peasant participation in the revolution"(Wolf, 1975,pp.385-86). The popular form of peasants' political actions that are studied most are the form of everyday resistance, that is "what people do that shows disgust, anger, indignation or opposition to what they regard as unjust, unfair, illegal claims on them by people in higher, more powerful class and status positions or instructions" (Kerkvleit, 2009, p. 233).

The idea of everyday resistance makes several of its parallels as it involves "little or no organisations; and the persons or institutions targeted by the resistance typically do not know, at least not immediately, what has been done at their expense (however, meagre it may be). The nasty, derogatory things peasants say or the jokes they crack about their landlords, employers, government officials, or the like behind their back" (Kerkvleit, 2009, p. 233) or what Scott (1985) says 'the ingenious ways of deceiving landlords and tax collectors. Such instances of everyday resistance serve as "precursors of open, confrontational, advocacy form of politics have occurred in numerous rural societies (Kerkvleit, 2009, p.234). This is better explained by Guha (1989) with the reference of colonial India, when peasant revolts inherently had its political connections. For example, Guha (1989) observes:

> When a peasant rose in revolt at any time or place under the *Raj*, he did
> so necessarily and explicitly in violation of a series of codes which defined
> his very existence as a member of that colonial power, and still largely

semi-feudal society. For his subalternity was materialized by the structure of property, institutionalized by law, sanctified by religion and made tolerable – and even desirable – by tradition. To rebel was indeed to destroy many of those familiar signs which he had learned to read and manipulate in order do to extract a meaning out of the harsh world around him and live with it. The risk in 'turning things upside down' under these conditions was indeed so great that he could hardly afford to engage in such a project in a state of absent-mindedness. (Guha, 1989, p.45).

In terms of establishing politics-peasantry relationship, Marx was probably the earliest and prominent social scientist who treated peasantry as a 'class', a social entity with a community of economic interests and political consciousness (see Marx [1950] 1987,pp. 331-337). He praised peasant involvement in the 1848 *coup d'état* (i.e 10 December 1848) which was the day of the *peasant insurrection* (italics was original). Indeed, 10 December was the *coup d'état* of the peasants which overthrew the existing government" (Marx [1950] 1987, p. 331).

Following Marx, Lenin proved it in the very first draft of the 'Program of Russian Social-Democrats (1884)' by the emancipation of Labour Group, "most serious attention was devoted to the peasant questions" (Lenin [1905] 1965, p.40]. He could see the peasants' potential for effective political action.

For Lenin, the question of the peasant movement was "vital not only in the theoretical but also in the most direct practical sense" (Lenin [1905] 1965, p.40) who later transformed the general slogans "to direct appeals by the revolutionary proletariat to the revolutionary peasantry" and he further urged, "The time has now come when the peasantry is coming forward as a conscious maker of a new way of life in Russia" (Lenin [1905] 1965, p.40).

One of the reasons of radicalization of peasants-politics relation was facilitated by the peasant (rural)-urban alliance. Such relations were resulted almost always from "an initiative by the urban partner...timing and circumstances for such an alliance are primarily determined by events in the national political arena" (Powel, 1971, p.2). Powel observes, "[T]he impulse for the alliance arises when an urban elite, engaged in a struggle for power at the centre perceives the need for a massive base of support and determinants that such a base is available in the peasantry in return for agrarian reform" (Powel, 1971, p.2). After a wide scope study of peasants' resistance in Southeast Asia, Scott long ago explicated the peasants-political interdependency as follows:

> There is little doubt that the rebellion would not have taken on the quite the dimensions of size and cohesiveness that it attained had it not been for the role of the communist party. While the precise role of the party does not materially affect the argument…the connection between official communist party and the peasantry is instructive (Scott, 1976, p.147).

Following the detailed review of contemporary peasant mobilizing discourses and practices, Philip McMichael (2008) concludes, "…peasant trajectories are conditioned by world, rather than national history" (McMichael, 2008, p.206). He urges us to locate the peasant movements into a larger-long term context. Having studied the peasant movements in the past and also at the Asia-Pacific regional level, Thapa (2005) summarizes, "Development of communism and its penetration in Asia is the root to originate peasant radicalism in Asian societies, and in turn, peasantry served as the source of mobilization for the communists" (Thapa, 2005,pp.29-30). This was quite clearly mentioned in its manifesto and in the first pamphlet released by the Communist Party of Nepal in 1949 (see Rawal, 2007).

At the same time, state or the landlords always denied the peasants' issues as genuine economic and social issues, issues that are related to the livelihood and dignity of the peasants, rather accused them of being a 'political' and their revolt as 'political move'. The government of Burma published a report on peasants' insurgencies of 1930s, which concluded, "…the rebellion must be regarded as primary political rather than economic in origin" (Scott, 1976, p.152). This reflects the universal character of peasants' insurgencies across time and space. For example, in the case of Chitang insurgency in Eastern hills of Nepal, the then Prime Minster in an interview with us argued that peasants from Chintag were setting up a war against the state. Hence, the response from the state or the landed class elite also has some universal characteristics.

The peasant mobilization, in the contemporary world, thus "...reaches beyond the daily round of survival on the land...transcends the conventional peasant politics, reframing its ontological concern...and reformulating the agrarian question in relation to development exigencies today" (McMichael, 2008, p.207). Across all time and space, every peasant movement we reviewed had its direct but a complex link to the politics. It can be generalized now that peasant movements constituted the political movements. This conclusion enables us to find a useful framework to analyse the case in hand; notwithstanding the peculiarity of all cases, each case has to be studied in its specific context, linking it to larger context. Underlying strategic relationship between the political parties and the peasants' groups and movements need to be explored and understood very carefully. It is noteworthy that peasant insurgencies in Nepal of any given period are unique in a sense that they do not have any parallel in contents and forms in Europe and even in Asia, including in precolonial India. Nevertheless, at the same time, some universality can never be ruled out.

Approaching Peasant Movements

An intricate relationship between peasantry and politics enables us to understand how peasantry can necessarily be a political factor; however, approaching and analysing such peasants' movement in context is always a complicated task.

Guha (1989), by reviewing the corpus of historical writings on peasant insurgencies during colonial India, encapsulate the discourses into three types as primary, secondary and tertiary (see Guha, 1989). They were useful in Indian context and possibly in other countries with colonial past; but do not fit quite well into Nepal's context as she never had direct colonial experiences in the past. Nevertheless, they can be helpful to draw inferences for our purpose.

Unlike 'colonial historiography that "...excludes the rebel as the conscious subjects of his own history and incorporates the latter as only a contingent element in another history with another subject"(Guha, 1989, p.77). He suggests to "... situate the uprising in a cultural and socio-economic context, analyse its causes, and draws on local records and contemporary accounts for evidence about its progress and eventual suppressions" (Guha, 1989, p.65); and, to acknowledge the insurgent "as the subject of his own history" (Guha, 1989, p.82). Here it would be relevant to refer Karl Marx, when he says, human beings "make their own history, but not of their own free will; not under circumstances they themselves have chosen but under the given and inherited circumstances with which they are directly confronted" (Marx 1950 [1852], p.146). Hence, the recognition of the peasants, the rebels, with their political agency and use of the rebels' narratives to avoid personal and hermeneutic biases provide a broader base for approaching the case.

Methodologically, O'Brien and Roseberry suggest to look at "the context [of inequality and contestations), the process [of struggle and the place of naturalized and oppositional historical images within it; and the production [of the historical images themselves]" (O'Brien and Roseberry, 1991,p.12). This is also required because "A subordinate movement needs to present itself in terms that its members and opponents can understand" ((O'Brien and Roseberry, 1991,p.13). Hence, voices and [counter] arguments of the alleged, opposition, state representatives of that time also need to be included to understand the larger context of the peasants' move.

Gellner, emphasising on collaboration between history and anthropology, puts it quite aptly in his Foreword in *Land, Lineage and State: A Study of Newar Society in Mediaeval Nepal.* (English Edition 2015. Social Science Baha/Himal Books) by Prayag Raj Sharma as follows:

...if our general aim is a more fine-grained and variegated understanding of South Asian history, we should be attempting to bring...anthropology and history into a proper relation, to find the right-way of using the present to understanding the past and the past to understand the present (Gellner, 2015, p.xiii).

This appears to be true not only for South Asian studies but also for any study on peasantry in any part of the world. Apropos of it, Wolf maintains:

> ...the enlargement of the scope of the peasant studies has had three important effects. First, it has brought about the notable convergence in the efforts of sociologists, anthropologists, political scientists, and of economic and social historians... the concern with the problems of peasantry has become one of the growth points of interdisciplinary comparative research, less through institutional organization than through convergent interest shared by a number of scholars (Wolf, 1975, p.386).

Another important dimension is the unraveling of the complexities between peasantry and politics. Kerkvleit further argues:

> Everyday politics is a fruitful realm of study in peasant studies. Much of the politics in peasant societies is of the quotidian sort. Hence, if one looks for politics only in the usual places and forms, as conventional political studies would direct, much would be missed about villagers' political thought and actions as well as relationship between political life in rural communities and the political system in which they are located (Kerkvleit, 2009, p.227).

Political issues permeate everyday life in the rural setting in particular (see Dhakal, 2012; Kerkvleit, 2009). "...creating and maintaining networks in order to have access to land, labour, money, and emergency assistance is a big part of people's everyday politics" (Kerkvleit,2009, pp.235-36) where the "mobilization and solidarity...occur within a particular structural and cultural context" (Galjart, 1976, p.3). My own work (Dhakal, 2012) has illustrated the politics in everyday life taking an example of eastern *Tarai* of Nepal. For such a ethnographic case study, as Yogesh Raj (2010), a historian, appositely puts that 'peasants' narratives' provide yet 'one more source material to work with' for social historians.

Peasants' Resistance and Political Mobilization in Nepal

Peasants' movements are often associated with exploitation and disobedience frame; however, Scott cautions (1976) that "growing exploitation of the peasantry may well be a necessary cause of rebellion, but it is far from a sufficient cause" (Scott, 1976,p.183). In this section, we endeavour to unpack some of the intricate, overt or concealed linkages between peasant uprising and political development and politicization of peasantry in Nepal. Wolford (2007) pertinently maintained that, "Micro-politics articulates with macro-politics in the sense that the strength of the individual movements draws on the ability of members to recognize and connect their particular conditions and political projects" (Wolford, 2007 quoted in McMichael, 2008, p.223).

This provides us a broader perspective to view Nepal's peasant movements in the past. Peasants' insurgencies in the past were against landlords, the large landholders, who also held political authority of some degree at the local level; hence, were also against the political establishment. As in the colonial India, the causes of such insurgencies of varied scales over different times and spaces were the exploitative nature of "tributary mode of production ...in which the primary producer, a cultivator is allowed access to the means of production, while tribute is exacted from him by political or military means...a ruling elite of surplus takers...at the apex of power system will be strongest when it controls some strategic elements in the process of production" (Wolf [1982],1997, pp.79-80). This fits into our context because Nepal itself was a 'tributary state' (see Regmi, 1976).

While trying to have a general overview of peasant insurgencies and the political mobilizations, we have focused on some fundamental questions as follows: (i) how well-thought were insurgencies and how well-thought were the planned executions after adequate consultation with the local peasants or their representatives?; (ii) what motivated peasants to take part in the insurgencies even by risking their lives and jeopardizing their livelihoods, and (iii) were the participating peasants conscious in order to understand the nature and patterns of peasant insurgencies of early years?. However, this section, peasants movements in retrospect, is based on the review of secondary sources, and hence, we may not get the satisfactory answers to all those questions.

The earliest connection of peasantry and politics is illustrated in the earlier documents of the political parties. For example, the first pamphlet of the Communist Party of Nepal (CPN), released on 24 April 1949 (12th Baisakh, 2006 B.S) had maintained that, among others, "[T]he peasants have to struggle for quenching their hunger, increasing the value of their farm products and their wages, and owning the land. So they also want liberation. Likewise, it also raised demands for the increased price of the crops produced by Nepalese peasants and increased the rate of their wages' (see Rawal, 2007, pp. 189-196). For them, the large landholders represented the 'exploiter class'. Hence, peasants were considered as a political force for the 'class struggle'; therefore, addressing their demands for ameloriating their downtrodden situation continually remained as one of the core agenda of all political movements since then.

The manifesto of the first Communist Party of Nepal (1954,pp.10-15) categorically emphasized and urged peasants to struggle for their rights and freedom. 'Total abolition of feudal autocratic regime' was its first agenda, and its fourth agenda says, "Elimination of all sorts of feudal exploitations including the landlord system (without compensation) and distribution of the land among peasants who cultivate the land, payment of the rural loans, and the provision of payments to the agricultural labourers sufficient for their living' (see Pushplal, 2055, Rawal, 2007).

The manifesto addressed the issues of the peasants and labourers. For instance, it wrote "Your freedom is possible only by means of a powerful struggle against the existing feudal system without a compromise… there is a need to initiate to make people's organization throughout the country under the leadership of the red flag, and launch struggle to fulfil the demand, i.e 'land to the tillers' set by the CPN" (Pushplal, 2055 BS,p. 79, Rawal, 2007, pp. 197-211).

The focus on peasant organization and urge for struggle was a long sustained feature of communist parties throughout the years. They are also well manifested in the early writings of Pushplal, the founder of communist party in Nepal (see Pushplala, 2055 B.S). Hence, after the establishment of the communist party, it got oriented towards the formation of 'people's organizations' and 'raising the struggle' though them (Mishra, 2058 BS,p.16, Rawal, 2007).

At the same time, Nepali Congress (NC), initially Nepal Democratic Congress and later Nepali National Congress, was more focused on political freedom and social reforms. However, it included agenda of 'problems of Nepali peasants and issues related to other economic life' since its first convention. Its manifesto of 1949 [1950], (Thapa, Poudel and Tiwari, 2057 BS, pp. 36-41) mentioned that:

> "...it is unjust to hold maximum land of the country by 5 percent landed elites making 95 percent people landless...NC is of the opinion that there should be a justifiable redistribution of land...managing a just land redistribution replacing the old unjust one is going to be the biggest economic revolution. Nepali Congress has to make people aware of such revolution...In Nepali Congress' view, the land held by big landlords and *Birta* holders should be distributed to laborious peasants and farm labourers. The economic condition of people is connected to the land. And, Nepali Congress being the representatives of all Nepalese, it assumes to be the only representative and well-wisher of peasants and farm labourers" [translation is mine].

After the end of Rana oligarchy, Nepali Congress strongly highlighted the issues of landlordism and *Birta* in favour of tenants and peasants. Eventually, *Birta* was abolished by Nepali Congress government in 1959. However, during Matrika Pd Koirala's three terms' premiership between 1951 and 1954, he not only dismissed the peasants issues but also supressed their voices (Tuladhar, 2046 BS, p. 35).

Tuladhar (2046 B.S), a political scientist and Thapa (2001), a historian, succinctly sketch the complex relationship between peasant movement and political movement in Nepal during the decade of 1950s (see also Rawal, 2007, KC, 1999). Tuladhar argues that it would be not only 'incomplete' but also 'biased' if one ignores the background of peasants' struggle while 'portraying the political, social, and economic aspects' of Nepal (Tuladhar, 2046 BS, p. 33). Hence, with the introduction of democracy in the country in 1950, growing activities of political parties also invariably surfaced the issues of peasants from different perspectives. One of the reasons of the growing concern of political parties in the peasantry was that the peasants constituted the majority of the voters and the elections were to be held sooner. Therefore, political parties were inclined to maintain a close alliance with the peasants. Even a party that represented the 'landed class' such as *Nepal Rastrabadi Gorkha Parishad,* was engaged in movement for lowering the rate of rent, of the interest rate, and waiving of old debts (Tuladhar, 2046 BS, p. 36), let alone NC and CPN.

Perhaps, B.P Koirala was aware and worried about Nepali Congress' inclination towards those landlords, and he urged that, "Nepali Congress should clearly spell out that our organization does good to the peasants... [there is] a fear that capitalist and feudal minded landlords might dominate our organization...we have to replace the old feudal land tenure system by such a system under which the peasants working in the farm will get the land rights" (Tuladhar, 2046 BS, p. 44).

The then Minister of Food, Bharatmani Sharma of Nepali Congress, responding to the peasants upsurge in Rupandehi of West Nepal in 2008 B.S (1951) maintained that, "...after the constitution is promulgated by *bidhan parishad* [constituent assembly], the land belongs to the tillers, therefore, vote for the true representatives; Nepali Congress does not want to please 5% at the cost of disappointment of 95%"(See Tuladhar, 2046BS). (Translation was mine; italics added).

Hence, keeping the forthcoming election in mind, every single party was trying to get support from the peasants; however, how sympathetic they were towards peasant's issues was not clear. Peasants were seen as "vote banks". It is worth noting that, when the election seemed to be uncertain during 2012 B.S (1955), parties became less interested in peasants' issues, and consequently peasant movements were slackened (see Tuladhar, 2046B.S). Tuladhar even suspected that the 'election was held when the peasant movements were in moribund situation' (Tuladhar 2046 BS, p. 43; see also KC, 1999).

Tuladhar (2046 BS) observed that NC had already adopted 'mixed-economy' during its first tenure in the coalition government (2008-2009

BS) and it denied the slogan it raised, 'the land to the tillers' during its incipient stage. First elected government of Nepali Congress under the premiership of Bishweswor Prasad Koirala (2016-2017 BS) declared a *Birta* Abolition Act (2016 BS) but eventually it turned the same *Birta* owners as the land owners (Tuladhar, 2046 BS) which did not help the tillers and actual tenants. It was not a redistribution of the land where peasants could have benefitted.

Along with political factions, peasants also got divided accordingly. *Kishan Panchyat* formed in 2010 B.S in Bara, Parsa, Rautahat was close to Nepali Congress whereas left inclined *Akhil Nepal Kishan Sangh* was established in the same year (see Tuladhar, 2046 BS, Jyapu, 2046 BS). Hence, from the very beginning, the growth of the peasant movements in Nepal was closely linked with the political parties. Communist parties were more engaged with peasant's issues and activities, and even the then government used to supress the peasants' uprising alleging them of being communists' (Tuladhar, 2046 BS, p.39).

Peasant revolts and uprisings were 'brutally supressed' during the decade of 1950 as the landlords had 'strong influence and decisive roles' in forming and dissolving the government in those interim years (Tuladhar, 2046 B.S, p.42). To oppose the growing peasants' insurgencies, the landlords also organized to protest the movement and protect their interest. Major protests included *Birtwal Sangh* Kathmandu (2008 B.S), *Bhumijibi Sang* (2010 B.S), and *Janahit Sangh* (2016 B.S) (see Thapa, 2001, Baral, 2058 BS, Tuladhar, 2046 BS).

After the seizure of power by the king Mahendra after dissolving the elected government in 1961, all political parties and their 'peoples organizations' were banned. This imposed a restriction on peasants' movement (Tuladhar, 2046 BS,p. 39, Jyapu, 2061,p.94). Notwithstanding this fact, he continued the reforms that began during the elected government's period to impress the general public. He declared the land reform in 1964. This was, as Powel (1971) describes elsewhere, "...a new elite seeking legitimacy...to mobilize a peasant base of support...to win the support of specific groups, to create or restore political stability; to legitimize their own political positions, or to create what they consider to be democracy" (Powel, 1971, pp.2-3).

Peasant's movements of the 1960s and 1970s are, in many instances, attributed to the failure of the land reform program of 1964 which raised the hopes of peasants and tenants. Had the land reform program fulfilled some of its promises, incidents like Chintang would not have taken place in the same manner as they occurred in 1979. Declining productivity and denial of ownership of land where redistribution remained as a fairy tale led to growing frustration and dissatisfaction among the peasants. Consequently, youth leaders organized in communist parties entered into the village with *Red Books* of Mao Tse-Tung and tactically 'exploited' the revolutionary potential of the peasantry, which, however, was largely denied by orthodox Marxism (see Roy, 2006). However, it, as many sporadic peasants movements of those decades, was bound to fail as it raised the issues of national scale, the question of changing the system and altering the existing power structure but the 'action' was too localized without adequate preparation.

Only in later 1970s and early 1980s, with burgeoning of the communist parties in the different parts of the country; peasants uprising began to come on the surface again. But, the nature and the structure of the political mobilization of peasants had not changed much since the beginning. It was, as Wolfs (1975) observed, "...peasant as clients in networks of political patronage operating on the level of the region and the nation (Wolf, 1975, pp. 385-86). Nepal's case was not unique. They can be compared and to some extent generalized. For example, Galjart (1976), based on his Chilean case study, writes:

> "...the problem of peasant mobilization and solidarity cannot be viewed exclusively in terms of the desires, aims and insights of the peasant themselves. It must be viewed within the larger framework...political leaders and movement...seek to influence the peasants' aspirations and actions" (Galjart, 1976, p.5).

Galjart's observations of Chilean case well resembles with the peasant mobilisation in Nepal of the same decade. In Sindhupalchok, tenants' rebel against *Guthi* and exploitative nature of land owners (2035 BS); agricultural labourers' movement for raising the wage rates and claiming the respects in many villages of Dhanusha in 2036 BS; and movement by jute farmers in Morang, Sunsari and Jhapa to increase the price of the produce, access to market and to increase the quota' were some of those movements where communist party got involved in (see Dhakal, 2013, Dhakal et al., 2000, Baral 2058 BS; Mishra, 2058 BS, NRPU, 2064 BS, Thapa, 2001).

Political-economic changes in late 1980s and 1990s influenced and impacted the peasant movements to give them a different turn which was deviated from its usual path in the past. As a limitation, we discuss the period prior to 1990s only.

Hence, as elsewhere, Nepal's peasants' movements also experienced the similar relationship with the politics which burgeoned along with the political process and movements of the left politics in particular. This also means that peasant movements were often led and represented by political leaders (belonging to landed class). Such pseudo representation could have been one of the reasons of the eventual failure of the peasant movements.

Peasant and Anthropological Concern: Locating the Self with Epistemological Questions

Nepali anthropology is floundering with the fundamental questions regarding the peasant studies in the country. We have been echoing for long that Nepal is an agrarian society. But questions such as "who constituted such a society"? and "how much is it explored and explained"? by anthropologists are still largely unanswered. We do not have an unanimous or a widely accepted definition of 'peasant' in Nepali context in general and in anthropological studies in particular. A generic Nepali term 'Kishan' for 'peasant' does not adequately describe 'peasant' in a particular anthropological context. Ethnographic works in Nepal by native or foreign anthropologists touch upon 'peasant issues' in different aspects and degrees. However, there have been relatively fewer works done on 'Nepali peasants' and 'peasant movements', leaving a significant space for the future anthropologists. Anthropological works that unravel the intricate relationships between 'peasants' and 'politics' in the given historical context are the 'prime locations' to delve into the peasant issues in Nepal. Efforts have to be made to establish the 'peasant' a robust 'anthropological type' rather than simply a 'disruptive category'.

This descriptive paper based on review of secondary literature has not been able to answer all the questions adequately. This is rather a 'provocation' or a 'wake-up call' for our own community to indicate that the time has come to orient our work towards more urgent questions of peasant studies.

It is significant to note that peasants at present are not what they were in the past. Therefore, a conceptual and methodological innovation is equally important. As Kearney (1996) in his '*Reconceptualising the Peasantry*' argues:

> "The shift from an anthropology of peasants as unitary objects to peasants as complex subjects marks a maturation of peasants studies whereby the very idea of peasants is transcended such that its continued invocation serves only to constrain the anthropology of communities formerly know as rural and peasant" (Kearney, 1996, p.7).

Hence, Kearney (1996) provokes us to question the relevancy of the earlier studies on 'peasants' (Firth, 1952, Redfield, 1960, Wolf, 1955, 1969 & 1975, Dalton, 1972, Nurge, 1971, Galjart, 1976, Ortiz, 1987, Shanin, 1990) in our particular context in conceptualizing, characterizing and explaining the peasants. In such case, we better begin with dealing a question, that is, 'should we need to develop our own concept and methods to fit more to our own unique social historical context'?.

Approaches and methods to study the linkages between peasants and politics, as discussed by Powel (1971), Wolf (1975 & 1982), Scott (1976), Guha (1989), O'Brien and Roseberry (1991), Wolford (2007), Kerkvleit (2009) and Gellner (2015), are appropriate points of departure to study 'Nepali peasants' and their 'political mobilization' and we can develop further an approach that adequately captures the political historical context that can provide ethnographic ground to study such linkages.

Finally, I propose some seemingly mundane questions to engage us in the field. These would comprise: (i) is peasant a class or a social category?;

(ii) what position they would have in a new identity politics?; (iii) how was state-peasant relation defined in the given process of the rise of modern nation-state?, and (v) do 'differentiation' and 'typification of peasants' still hold anthropological signification?. Only by answering these questions, we can establish that the study of peasants is still a relevant field of study in our context. I end this paper here with a hope that we need to see more anthropological engagement in the study of 'peasants' and 'peasants' politics' in the years to come.

Acknowledgements

The substantial portion of this paper was drafted during my stay at Graduate School of Asian and African Area Studies (ASAFAS), *Kyoto* University, as a Research Fellow during the period of February-July, 2016. I am grateful to Prof. Tatsuro Fujikura of ASAFAS for his academic support while writing this paper. I express my sincere thanks to Prof. Dilli Ram Dahal who provided me useful remarks upon reading the preliminary draft of the paper. I am, in particular, thankful to Dr. Dambar Chemjong for his critical comments on the paper during the seminar jointly organized by the Central Department of Anthropolgy, Tribhuvan University and Community Self-Relaince Cetnre (CSRC). I am also thankful to Prof. David Seddon and other participants of the conference for providing useful feedback.

References

Baral, T. (2058 B.S). *Atitka janasangharsha (Peoples' movements of the past)*. Dhankuta: Byanjana, Kshitiz and Sindhu Baral.

CPN-ML (2017 B.S). *Barga shangharsa* (the class struggle).Kathmandu.

Dahal, D. R. (1981). Concept of economy in a peasant society: A case study of the Athpahariyas of east Nepal" *Contribution to Nepalese Studies* Vol. VIII no.2. June 1981. CNAS. Pp. 55-71.

Dalton, G. (1972). Peasantries in anthropology and history. *Current Anthropology* 13:385-416.

Dhakal, S.(2011). *Land tenure and agrarian reforms in Nepal*. Kathmandu:CSRC.

Dhakal, S. (2007). '*Haruwas*, the unfree agricultural laborer: A case study from eastern *Tarai*'. In *Contribution to Nepalese Studies*. Vol 34, No.2. Journal of Center for Nepal and Asian Studies. Tribhuvan University.

Firth, R. (1951). *The elements of social organization*. London: Watts.

Galjasrt, B.F. (1976). *Peasant mobilization and solidarity*. Assen/Amsterdam: Van Gorcum.

Gellner, D. (2015).Foreword. In Pragayraj Sharma's *Land, lineage and state: A study of Newar society in mediaeval Nepal*. Kathmandu: Social Science Baha, Himal Books.

Guha, R. (1989). The prose of counter-insurgency. In R. Guha and G. C. Spivak (Eds.) *Selected subaltern studies*. Oxford: Oxford University Press. Pp. 45-84.

Guha, R. (1983). *Elementary aspects of peasant insurgency in colonial India*. [Oxford India Paperback 1992]. Delhi: Oxford University Press.

Johnson, K. Wisner,B. and O'Keefe,P. (2005). Thesis on peasantry revisited. *Antipode*. Vol 37 (Issue 7). Pp 944-955.

Jyapu, R. (2046 BS). Bara *ra Rautahat Jillako kishan andolanko lekha Jokha* (An assessment of the peasant uprising in Bara and Rautahat) In *Jhilko* 12. Vol 2. Pp. 18-32.

Karki, A. K.(2002). Movements from below: Land rights movement in Nepal. *Inter-Asia Cultural Studies (Volume 3, Number 2, August)*.

KC, S. (2065 BS). *Nepalma communist andolanko itihash (in Nepali). [Lit trans. History of communist movement in Nepal]*. Vol. 3. Kathmandu: Viddyarthi Pustak Bhandar.

Kearney, M. (1996). *Reconceptualising the peasantry: Anthropology in global perspective*. Colorado: Westview Press.

Kerkvliet, B. J. T. (2009). Everyday politics in peasant societies (and ours). *Journal of Peasant Studies*. Vol. 36. No 1. Pp 227-243.

Keyes, C. F. (1966). *Peasant and nation: A Thai-Lao village in Thai state*. A Thesis Presented to the Faculty of the Graduate School of Cornell University for the Degree of Doctor of Philosophy.

Lenin, V.I. (1905). The proletariat and the peasantry. *Lenin Collected Works*. Moscow: Progress Publishers. Vol. 10, Pp.40-43.

Marx, K. (1987). Peasantry as a class. In Teodor Shannin (Ed.). *Peasants and peasant societies: Selected readings*. New York: Basil Blackwell. Pp. 331-337. Original excerpted from Karl Marx , Marx, K. (1950). The class struggle in France 1848. In K. Marx and F. Engels. *Selected Works*, Vol. 1, Foreign Language Publishing House. [Original 1852].

Marx, K. (1950). 'The eighteenth Brumaire of Louis Bonaparte. In K. Marx and F. Engels *Selected Works,* Vol. 1, Foreign Language Publishing House. [Original 1852].

McMichael, P. (2008). Peasants make their own history, but not just as they please. In *Journal of Agrarian Change.* Vol 8(2). Pp. 205-228.

Mishra, B. (2058 BS). *Jhapa andolan dekhi e ma le bibhajan samm (From Jhapa uprising to UML split).* Kathmandu: Vidhyarthi Pustak Bhandar.

Nurge, E. (1971). A peasant is as a peasant does. *American Anthropologists.* New Series. Vol.73 No.5. Pp. 1259-1261.

Novyaya Zhizn (November 2, 1905.). The proletariat and the peasantry. *Lenin Collected Works* (1965). Moscow: Progress Publishers, Volume 10, pp. 40-43).

O'Brian, J. and Roseberry, W. (Eds.) (1991). *Golden ages, dark ages: Imagining the past in anthropology and history.* Berkley: University of California Press.

Ortiz, S.(1987). Peasant culture, peasant economy. In Teodor Shannin (Ed.). *Peasants and peasant societies: Selected readings.* New York: Basil Blackwell. Pp. 300-303.

Poudyal, D.(2016). Ethnic identity politics in Nepal: Liberation from or restoration of elite interest? *Asian Ethnicity.* DOI:10.1080/14631369.2016 .1179567.

Powel, J. D. (1971). *Political mobilization of the Venezuelan peasant.* Cambridge: Harvard University Press.

Pushplal (2055 BS). *Pusplal: Chaniyaka rachanaharu bhag 4* [Pushplal: Selected works, Part 4]. Kathmandu: Pushplal Smriti Pratisthan.

Raj, Y. (2010). *History and mindscapes: A memory of the peasants' movements of Nepal.* Kathmandu: Martin Chautari.

Rawal, B. (2007). *The communists in Nepal: Origin and development.* Kathmandu: Achham-Kathmandu Contact Forum, Communist Party of Nepal (UML).

Redfield, R. (1960). *The little community and peasant society and culture.* Chicago: The University of Chicago Press.

Regmi, M.C. (1976). *Land ownership in Nepal.* Berkley: University of California Press.

Regmi, M.C.(1977). *Land tenure and taxation in Nepal.* Kathmandu: RatnaPustakBhandar.

Roy, H.(2006). *Peasants in Marxism.* New Delhi: MANAK Publications Pvt. Ltd.

Scott, J. C. (1985). *Weapons of the weak: Everyday form of peasants resistance.* New Heaven: Yale University Press.

Scott, J. C. (1976). *The moral economy of the peasant: Rebellion and subsistence in Southeast Asia.*

Seddon, D; P. Blaikie & J. Cameroon (1979). *Peasants and workers in Nepal.* Delhi: Adroit Pulishers.

Shanin, T. (1988). *Defining peasants: Essays concerning rural societies and peasants studies.* Basil Blackwell.

Sharma, P. R. (2015) *Land, lineage and state: A study of Newar society in mediaeval Nepal.* (English Edition). Kathmandu: Social Science Baha/Himal Books.

Sisson, R. (1969). Peasant movement and political mobilization: The *Jats* of Spencer. Jonathan (1996).'Peasants'. In Barnard and Spencer (Eds.) *Encyclopedia of social and cultural anthropology.*London: Routledge.Pp. 418-419.

Thapa, G., Poudel, P. & Tiwari,P. (2067 B.S.). *Nepali Congress: Aitihashik dastabejharu* (Nepali Congress: Historical Documents). Kathmandu: Public Policy Pathshala Pvt. Ltd.

Thapa, K. B. (2046 BS.19aun) *Satabdiko purbardhatira nepal ma Kkshan samasya* (Peasant's problems in Nepal around the first-half of 19th century).). In *Jhilko* 12. Vol 2. Pp. 3-8.

Thapa, S. (2005). Conceptual framework to study peasant society and economy'. In *Voice of History.* Vol XVII-XX (1) Pp. 29-50.

Thapa, S. (2001).*Peasant insurgencies in Nepal:1951-1960.* Bhaktapur: Nirmala K.C.

Tuladhar, B.(2046 B.S.). *Nepalko rajnitik bikasma kisan andolanko bhumika: Dalgat rajnitiko ek dasak (2007-2017)* (Roles of peasant insurgencies in the political development of Nepal: A decade of party politics 1950-1960). In *Jhulko* 12. Vol 2. Pp. 33-45.

Watters, R.F.(1994). *Poverty and peasantry in Peru's southern Andes 1963-1990.* Hound mills: Macmillan.

Wolf, E. (2001).*Pathways to power: Building an anthropology of the modern world.* Berkeley: University of California Press.

Wolf, E. (1997) [1982]. *Europe and the people without history.* California: University of California Press.

Wolf, E. R. (1975). Peasant and political mobilization: Introduction. In *Comparative Studies in Society and History,* Vol.17, No 4. Pp. 385-388.

Wolf, E. (1966). *Peasant.* New York: Prentice-Hall.

Wolf, E. (1955). "Types of Latin American peasantry: A preliminary discussion". *American Anthropologist* 57(3): 452-471).

Wolford (2007). *Quoted in McMichael* 2008. P.223.

Resistance, Rebellion, Revolt and Revolution: A Historical/Anthropological Consideration of Peasant Movements and Other Forms of Rural Unrest in Nepal

DAVID SEDDON

Introduction

In his recent review of 'political violence in Nepal', Whelpton notes that, in 1850, Jung Bahadur Rana told Orfeur Cavenagh that 'although revolutions often occurred (in Nepal), yet the country as a whole did not suffer more from such disturbances than England would from a change of Ministry; neither the army nor the peasantry taking any part in the disputes, and submitting without a murmur to the dictates of whichever party might emerge the victors' (Whelpton, 2013, p. 27). The notion that 'revolutions' were about struggles between factions at the national level, without the involvement of the army or the peasantry, is one that tends to be perpetuated in the numerous 'histories of the Ranas' (mainly by Ranas) which focus on palace and court intrigue, and on factional divisions within the ruling class.

Careful study suggests, however, that in reality, the regime over which Jung Bahadur and his Rana successors presided for many years (1846-1951) was subject to various forms of rural unrest in which not only the peasantry but also the army was involved. Furthermore, the previous 'pre-Rana' or 'post-war'[1] regime (1816-1846) also experienced rebellion and revolt even after the 'unification' of the Gorkhali State; while the process of 'unification' itself involved protracted rural conflict, including various forms of resistance, arising from the reluctance of many of the petty states and tribal chiefdoms across the western and the eastern hills to recognize or to submit to the authority and rule of the new Gorkhali state.

In this chapter, I examine the history of rural unrest over a period of some two centuries, firstly (in Part One), in the context of the broader history of the 'unification' and development of the Gorkhali State, initially under the Shah rulers of what was formerly one of several petty hill states (Gorkha) and subsequently under the dynasty of hereditary prime ministers, the Ranas. Then (in Part Two), I shall consider the more recent period following the overthrow of the Ranas, from 1951 up to the 1990s. We hope one day to be able to develop this further, taking the discussion up to the present day.

We will also explore the differences and the relationships between various forms of rural unrest, including: resistance to incorporation within the state; rebellion against the authority and jurisdiction of the state; unrest, uprisings and revolts, as well as ideologically motivated movements aiming to challenge and even overthrow the state to bring about a radical transformation of economy and society. Our main focus, however, will be on peasant movements and peasant revolt.

Wolf has defined 'peasants' as rural dwellers who broadly are subject to 'the dictates of a superordinate state' (Wolf, 1973, p.xviii). Others have distinguished 'peasants', who are by definition generally subordinate to the state, from 'tribal' communities, that live effectively beyond 'the pale', and are generally not subject to state control (Gellner, 1969, Seddon, 1981). However, many pre-capitalist agrarian states were able to maintain

---

[1] The period after the 1814-16 Anglo-Nepali war.

effective control over the peasantry only under certain circumstances; and the varying limits to that control defined, at any one time, the physical boundaries between those areas where the peasantry was subject to state control and those areas where they were not, and also the 'marginal areas' between these two, where 'bandits' and outlaws often operated[2].

Even in modern states, the peasantry may also resist specific dictates of, or interventions by, the state, by a variety of means, including the deployment of what James Scott has called 'the weapons of the weak' (Scott, 1985), which may involve negotiation with the authorities, fleeing to escape the jurisdiction of the authorities, or, where these actions or reactions are not possible or fail, even revolt (see Wolf, 1973). Distinct, in my view, from peasant uprisings or revolts – which are usually responses of 'the last resort' to particular circumstances in which a material or moral contract is felt to have been broken[3] – are rebellions, which are generally led by local notables, and which challenge or refuse the authority of a particular regime, or even of the state more generally. These may, in fact, overlap with 'elite politics', where the rebellion effectively threatens the regime, or may have the appearance of 'ethnic unrest' where a rebellion involves rural elites and their followers among the peasantry who identify themselves as a distinctive tribal or ethnic group.

Whelpton himself distinguishes four main types of what he calls 'politically motivated violence' in the case of Nepal: conflicts within the political elites, ethnic and regional resistance, agrarian revolt, and 'crusades'. He argues that the first of these coincides pretty much with Jung Bahadur's own view 'that a successful replacement of the top power holders was always an elite affair, and that the mass of the population meekly accepted the dictates of the victor' – commenting that this was 'of

---

[2] For example, the 'water margins' of the Huainan region of mediaeval China, where the outlaw Song Jiang and his 36 companions were active during the 11th century CE, or the areas of '*siba*' (dissidence) in pre-colonial Morocco, where various forms of rural unrest, including rebellions against the *makhzen* (the state) and examples of social banditry, have been identified (Seddon, 1978).

[3] As when rents, taxes or consumer prices rise 'unreasonably' (see Walton & Seddon 1994 with reference to urban protest).

course, an oversimplified view but not too far out of line with the situation in the mid-nineteenth century' (Whelpton, 2013, p.28).

It could be argued, I suggest, that this 'situation' continued through into the early part of the 20th century and, indeed, thereafter, up to the present time, in so far as politics remains a largely elite affair, despite the claim that Nepal is (at least since the Jana Andolan of 1990 and subsequent revision of the Constitution) a democracy. Elite politics is often characterized, as has previously been suggested, as 'ethnic' or 'regional' politics in which the claim is made, generally by members of a disaffected or marginalized elite, that they represent a disadvantaged or oppressed ethnic (e.g *Limbu*) or regional (e.g *Madheshi*) population.

The second, he associates particularly with the circumstances of what he calls 'the original Gorkha conquest', but also suggests that 'there were, however, rather more instances of ethnically based revolts in the intervening years than is sometimes perceived' (p.28). The extent to which a rural uprising or movement may be identified as 'ethnic' depends, I suggest, on a number of factors, including the characterization of the uprising or movement both by outsiders and by the participants themselves. But uprisings and movements identified (by the participants and/or outside commentators) as 'ethnic' did not end in the late 18th and early 19th century CE; they have been a feature of rural conflict in both the 20th and 21st centuries CE, as we shall see.

As regards ' agrarian revolt', Whelpton comments that 'tensions between the peasantry and those claiming a share of their produce as rent or taxation could result in violence and, where both cultivator and landlord were from the same ethnic group, this could be seen as straightforward class conflict. However, in particular areas, matters were more complex, as the agrarian divide coincided with an ethnic one' (Whelpton, 2013, p.28) - the point made above. It may also be difficult in reality to distinguish or at least to disentangle peasant revolt from other forms of local and even national politics.

Finally, what he terms 'crusades' are, he suggests, 'rebellions driven by religion or ideology, ranging from the varieties of liberal and Marxist value-systems behind *sat sal*, the Maoist's People's War, and other, less successful uprisings in recent times, to the Lakhan Thapa revolt against Jung Bahadur analysed in detail by Lecomte-Tilouine (2003)'[4]. I am not, myself, comfortable with the term 'crusades', but would argue in any case that the extent to which various rural uprisings had an ideological (as opposed to a practical or pragmatic) objective or mission, remains something to be explored empirically, before they can be classified in a particular way.

Struggles for land rights, for example, are usually not *just* related to 'tensions' or class struggle between the peasantry and landlords and focused on access to land as the means of production – although they may be that. They are often also about 'rights' to land and, even beyond that, about a vision of a fairer society – in other words, a political ideology. Indeed, Whelpton acknowledges that 'one does not have to be a Marxist to acknowledge that ideology generally marches with material interests, so this factor usually operates in tandem with one or more of the other three' (2013, p.29).

He also comments that it is not always possible to distinguish between political violence and violence applied solely for private gain or straightforward criminality, and notes that if the distinction were 'hard and fast' there would be no call for what he calls 'an intermediate category' like 'social banditry' – a term coined by Eric Hobsbawm (1959). As we have already remarked, however, in early state formations, the distinction between those areas where dissidence prevailed and those where the state was nominally at least in control was often unclear, as the 'reach' of the state would wax and wane over time. But there were also, often, marginal areas between the areas under control and those beyond the control of the state, where 'dissidence' predominated. It was here, pre-eminently, that one found what Hobsbawm identified as 'social bandits'.

Finally, Whelpton usefully draws attention to the fact that not all of these categories involve purely indigenous or local actors but may be affected in various ways by 'external forces'.

---

[4] Also discussed elsewhere by Marie Lecomte-Tilouine (Lecomte-Tilouine 2000 and 2013).

PART ONE: 1750-1950

The Rise of Gorkha

In the 18th century, population was sparse, with various modes of pre-capitalist production predominating – hunting and gathering from the wild, pastoralism, and shifting agriculture for the most part; settled agriculture, in so far as it existed, was largely reliant on rainfall and irrigation was poorly developed in most of the petty hill states of what would eventually become Nepal (Regmi, 1999, p.18), and so the agrarian economy was not a reliable basis on which to rely for tax revenues from surplus production. More important for many hill rulers were other resources, including minerals (notably iron and copper) which were exported to India, and control over long distance trade[5], north to Tibet and south to India (pp.19-20).

It seems that Gorkha was particularly poor, 'being cut off from any direct communication with either the low country (the *Tarai*) or Tibet, and having no mines, nor other production as basis for commerce' (Regmi, 1999, p.5). Its expansionist ambitions, however, appear to date back as early as the reign of Narbhupal Shah (1716-1743). An attempt by Gorkha in 1737 to seize the trading centre of Nuwakot (on the borders of the Kathmandu Valley) from Kathmandu was not successful, but in 1744, its continued efforts were rewarded and the profitable trade that passed through Nuwakot came under Gorkha's control. It seems that, in the 1740s, Prithivi Narayan Shah established close relations with Indian arms manufacturers and dealers, and brought in technicians to assist with the local production of weapons (Regmi, 1999, p.6). This, together with the martial qualities of the men of Gorkha, established a decisive difference between Gorkha and its neighbouring states and helps explain its subsequent rise to power.

---

[5] The importance of control over long-distance trade for the maintenance of state power has been noted in a number of cases of early pre-capitalist formations, in Asia and elsewhere (cf Lacoste 1974, Murray 1975, Seddon 1981, Terray 1974).

Whelpton devotes several pages to the early 'unification' of Nepal during the late 18th century and discusses both initial resistance to Gorkha expansion and annexation, and also later rebellions against the new Gokhali state, the most persistent of which were in the eastern hills (2013, pp. 29-32). He remarks that 'Prithivi Narayan Shah's conquest was facilitated because, as the first hill ruler to import firearms, he had a technical advantage over many adversaries, including the *Rais* and *Limbus* of the eastern hills, who still fought with swords and arrows. During the 1774 campaign in this region, Gorkhali forces were easily able to take the *Tarai* capitals of the Sen rulers and also to over-run large tracts of hill country, but they then faced seven years of continuing resistance, probably on a guerrilla pattern' (Whelpton, 2013, p. 30).

By 1748, Sindhupalchowk and Kabre Palanchowk and other areas to the east of Kathmandu were subjugated, and the passes of Kuti and Kerung, which constituted the main routes through which flowed trade between the Kathmandu Valley and Tibet, also came under Gorkhali control. The conquest of the kingdom of Makwanpur in September 1762 gained the expanding Gorkhali state access to areas of the eastern *Tarai* through which the kingdoms of the Valley had maintained contact with India. The conquest of Kirtipur provided a foothold in the Valley itself, and in 1768, the kingdom of Kathmandu was overcome; a year later, Prithivi Narayan Shah also controlled Patan and Bhadgaon (Bakhtapur). By 1775, at the death of Prithivi Narayan Shah, the expanded Gorkhali kingdom, with its capital now in Kathmandu, included the whole of the eastern *Tarai*, the eastern and central inner *Tarai,* and the eastern hills up to the Tista River bordering Sikkim, and a small part of the western hill region, including Jajarkot.

The Eastern Hills

At the time of the rise of Gorkha in the mid-18th century, the *Kiranti* peoples of the eastern hills enjoyed a high degree of political autonomy, alternating their allegiance between the Malla kings of the Kathmandu Valley and the king of Sikkim. The emerging Gorkhali state expanded into the eastern hills only in the mid-1770s. The Battle of Chainpur, in which the Gorkhali forces were pitted against the armies of Sikkim, took place, for example, in 1776. After its victory there, the Gorkhali state began to consolidate its control of eastern Nepal, in two main phases: firstly, the ethnic groups or 'tribes' of 'near' (*wallo*) and 'middle' (*majh*)[6] Kirant, secondly those of 'further' or 'far' (*pallo*) Kirant, known as *Limbuwan* (or *Limbu* territory).

The effective annexation of the eastern hills is attributed by Pradhan (1991) to the willingness of some locals to fight against their own people, to the absence of a united opposition and to the fact that the locals were armed mainly with swords and bows rather than firearms. It does seem highly likely that the expansion of the Gorkhali state was facilitated by its superior military strength, combined with the successful negotiation of strategic alliances: 'the expansion eastwards of the Hindu armies in the 1770s resulted in a series of alliances with potentially troublesome *Limbu* chiefs flanking them on all sides', based on the promise that the *Kirantis* would, in return for their support, be granted a measure of internal rule under their own chiefs and guaranteed their rights to 'ancestral lands' (Caplan, 1970, p. 4).

Even so, there was strong resistance, particularly in *pallo* Kirant, where some local *Limbu* and Sikkimese forces held out until 1786. Fitzpatrick remarks that 'despite partial control of *Limbuwan* from about 1782, and more substantial control from 1786, political, cultural and economic resistance continued in various forms through to the fall of the Rana regime in 1951, and elements of this resistance could be said to exist still today' (2011, p.34)[7]. The political culture of the far eastern hill peoples,

---

[6] Today's Bhojpur and Khotang districts were part of *majh* Kirant. Later, the region was known for the purposes of army recruitment as 'east no. 4'. The majority of the population were *Rais*, but other ethnic groups, including *Limbus*, *Gurungs*, *Yakkas*, *Tamangs* and *Magars*, as well as high caste *Brahmins* and *Chhetris* and so-called occupational castes or *dalits*, also lived there. *Limbus*, however, generally tended to live further east, in *pallo* Kirant (*Limbuwan*).

[7] Indeed, as we shall see later, there was significant rural conflict in this region during the 1950s, and later, in part as a result of the growth of 'identity politics' since 1990, and even in the last five years, as the idea of ethnic federalism (led by the Maoists) resulted in a resurgence of political activity in the far eastern hills

who identified themselves as '*Limbu*' and their territory as '*Limbuwan*', has remained exceptionally strong over the last two centuries. One of the reasons for this relates to their distinctive forms of political and social organization founded on communal forms of land tenure (*kipat*) and a high degree of egalitarianism.

The economic base of local societies in the eastern hills had previously involved a combination of an earlier hunting and gathering economy with pastoralism and 'slash and burn' agriculture, associated with distinctive forms of local community land tenure, rather than individual ownership. While land tenure and land ownership in most of Nepal was organised in the late 18th around the *raikar* system, in which individuals could effectively hold private title (even if ultimate property rights rested with the state), land in the eastern hills was largely organised under the *kipat* system, in which land was held communally and individuals had rights and access to it by virtue of their membership of a specific group and through allocation by a local headman or chief.

Waste land and un-cleared forest as well as cultivated land were all considered *kipat* by the locals, whereas, elsewhere in Nepal, various forms of state ownership and individual title predominated. *Kipat* was thus a pre-capitalist form of 'community land tenure' of a kind originally discussed by Karl Marx. Unlike *raikar, kipat* land could, in theory at least, not be sold or given to non-community members. However, prior to 1883, there was no specific restriction on the ability of local communities to give away their *kipat* land, and land thus alienated was often registered as *birta* by those receiving or buying it.

It is not actually clear what control 'pre-unification' chiefs and the rulers of petty states in *Kirant*, where they existed, had over the allocation of rights to land, or, indeed, whether their powers arose from local circumstances or from recognition by a higher-level ruler (e.g the king of Sikkim); but Caplan explains that 'the Shah rulers did not create *subbas*

---

among ethnic *Limbus* around the notion of a distinct and potentially autonomous '*Limbuwan*' – as indeed did the notion of a distinct state of '*Khumbuwan*' in Solu Khumbu.

among the *Limbus*. They only absorbed the traditional headmen into the administrative structure of the new state (Caplan, 1970, p. 28).

Caplan remarks that the *Limbus* of the eastern hills 'did not submit docilely to Gorkhali subjugation and were not hesitant to seek external support for their rebellions against Kathmandu's authority'. During the Nepal-China war of 1788-93, many of them actually joined the enemy. Such rebellions were eventually suppressed, and many *Limbus* fled their homeland out of fear of reprisals; but the government recognised the need to develop more amicable relations with the *Limbus*. Accordingly, it proclaimed an amnesty in 1795, 'calling on the fugitive Limbus to return to *pallo kirat* and guaranteeing their traditional customs and privileges' (Regmi, 1999, p. 51) – including, notably their *kipat* collective land-holding rights.

The Western Hills

By 1789, resistance to the process of 'unification' had been largely subdued, the *chaubise* and *baise* western hill states had all been annexed and the frontier in the west extended to the Mahakali River (Regmi, 1999, pp.7-8). Once annexed, and reduced in effect to vassal status, these subordinate states initially paid tribute to their overlords, the rulers of the emerging Gorkhali state. Indeed, the political formation that emerged through 'unification' was effectively a tributary state – an amalgamation of subordinate political formations rather than a unified state – and one whose ruling elite was to struggle to retain power well into the 19th century.

In the hills to the immediate west of the Kathmandu Valley, and near to the original petty state of Gorkha, it seems that the Gorkhali state was seen by the locals largely as either a continuation of what they had known previously or as something that brought about minimal change in their immediate circumstances. Many peasants in this region, notably *Gurungs* and *Magars*, were recruited from one state into the Gorkhali armies and deployed to support the pacification of others further to the west; the benefits of this were almost certainly significant and generally appreciated by those concerned, even if others may have resented it.

Local rulers were generally brought under Gorkhali jurisdiction and the authority of the emerging Gorkhali state, but as Regmi explains, 'in the western and far-eastern hill areas, Prithivi Narayan Shah generally followed the policy of subjugating principalities without actually annexing their territories' (Regmi, 1999, p.12). 'As a general rule, the territories of defeated *rajas* were... annexed only when they resisted the Gorkhali invaders to the last, or rebelled after once accepting their suzerainty' (Regmi, 1999, p.13). This form of indirect rule, enabled the local rulers to maintain their capacity to exploit their own peasantries and thus sustain their position as overlords, despite accepting a degree of political subordination.

Jajarkot was the first principality to accept the status of a vassal under the Gorkhalis in this manner. In a letter to the *raja* of Jajarkot in January 1769, Prithivi Narayan Shah wrote:

'We confirm your ancestral authority within your territory, including your authority to award capital punishment, upgrade or degrade castes, collect levies to finance the sacred thread investiture ceremonies and weddings of royal princes and princesses, and fees for the expiation of caste offences. We also confirm your authority to grant or confiscate *birta* lands and to collect judicial fines, escheats and fees for stamping weights and measures. You shall pay only Rs. 701 whenever a new king ascends our throne. When a new king ascends your throne, you shall have authority to collect customary payments from your people'.

This policy, says Regmi, was continued by the successors of Prithivi Narayan Shah when the western and far-western hill areas were subjugated from 1786 to 1789; although circumstances sometimes compelled the Gorkhali rulers to annex territories directly, as when they resisted or rebelled. For example, Kaski, which had initially been given vassal status, was later annexed when the *raja* rebelled – an example of elite politics, one might suggest. Some of the *rajas* displaced by the Gorkhali state were, however, later restored as vassals when this seemed expedient to the Gorkhalis.

The *raja* of Bajura, for example, was granted such status in 1791 in appreciation of the assistance rendered by him during Gorkha's wars with Achham, Doti and Jumla (Regmi, 1999, p.13). Also, many of those who at first fought against the Gorkhali state, later joined as part of its armed forces against its enemies. For example, Bhakti Thapa, who was later killed while fighting in the Gorkhali army against the English in 1815, had originally been a commander in the army of the *raja* of Lamjung and had been taken prisoner by Gorkha in 1792.

The Far Western Hills

In the far west, as Whelpton comments, 'Gorkha rule seems to have been felt as more of an alien imposition and particularly strong resistance was encountered in Jumla, which had previously enjoyed a certain primacy among the *baisi* states of the Karnali basin' (2013,p.32). At the time of its conquest by Gorkha, Jumla controlled a number of other petty states, including Mustang, which had come under its suzereignty around 1760. The Malla rulers of Jumla were in a position thereby to secure significant revenues both from within their own territory and from other subordinate or vassal states. As early as 1751, a massive sum of Rs. 9,000 was raised, through a special levy, to employ labourers, stone-workers, carpenters and architects in the construction of a royal palace, to engage *Brahmin* priests to perform the appropriate religious ceremonies to sanctify the new buildings, and for the purchase of cloth, cows and other items required for these ceremonies (Regmi, 1999, p.21).

It seems that, in the Karnali region as a whole, including Jumla, the rights of peasants (*chuni ryots*) on the *raikar* (private) land they cultivated had already developed into a form that permitted transactions in land through the payment of cash; also payment of land taxes to the state were generally in cash. Regmi speculates that 'the fact that most of these areas had once formed part of the Malla Kingdom of Jumla and thus, unlike the other principalities of the (western) hill regions, had been subject to a common polity may have had some bearing on this…' (Regmi, 1999, p.30). The *chuni ryots* of the Karnali region differed from the majority of the peasants in the western hill states where the term used was generally *mohi,* or tenant farmer. As Regmi remarks, 'the status distinction is obvious enough' – the *chuni ryots* had greater control over their own production and could even raise the money to pay taxes without personally working the land; they were small landowners, whereas the *mohi* was formally a tenant and often a sharecropper.

Jumla was perhaps the most powerful of the *baise* states, and withstood the onslaught of the Gorkhalis for two years before its capital, Chinasim, fell eventually in 1789. Acounts differ as to what then happened to its ruler, Shobhan Shahi; some say he was arrested and brought to Kathmandu, others that he escaped to Mugu, from where he mounted a revolt, which he lost, before mounting another revolt from Humla, which he again lost, before fleeing to Tibet, where he died (Adhikari, 1998,pp.55-59). Jumla was finally pacified in 1793 and it was then governed by a *subba* sent from Kathmandu (Shrestha-Schipper, 2013, p.260).

The 'Unified' Gorkhali State

Although Jumla thereafter accepted its subordinate status within the Gorkhali state, many Jumlis who were not willing to do so voted with their feet. Whelpton comments that Bishop (1990, pp.122-146) 'used revenue statistics to calculate that there was a substantial dip in population, which did not return to the pre-conquest level until 1930. Some other principalities in the far west continued to be restive for some time into the 19th century' (Whelpton, 2013, p. 32).

By the end of the century, however, the western *chaubisi* and far western *baisi* principalities had all been subjugated or annexed, and the frontier of the Gorkhali state extended to the Mahakali River in the west. By 1808, the Gorkha army had reached Kangra, across the Jamuna River. By this time, the Gorkhali state had more or less established its hegemony over a significant territory extending approximately 1,300 miles from the Sutlej River in the west to the Tista River in the east. But it continued to press for greater control over its territories in the *Tarai,* as these were also a significant source of state revenue. This eventually led to increased tension and eventually, in 1814, conflict with the British East India Company, which was also ambitious about controlling and administering territories in the Ganges plains adjacent to the southern territories of the Gorkhali state.

One might best characterise the Gorkhali state as 'a tributary state'[8], accruing wealth essentially through its power, authority and jurisdiction over the subordinate states, which had vassal status and gave 'tribute' to Gorkha; and also increasingly by its own control over an increasing number of trade routes, from which it derived revenues, over the natural resources of the forests and, increasingly important, over an expanding cultivated area of settled farmers and peasants, from whom it extracted surplus in the form of rent and or tax. This could be referred to as a form of what Marx termed 'the Asiatic mode of production'.

When, in 1792, Gorkha's expansionism towards the north and its aggressive policy towards Tibet led to the intervention of the Chinese army, the Chinese called on the relatively newly subjugated peoples of the central and eastern hills to join them, and to rise up in revolt against their new masters. It is not clear to what extent they were calling on local rulers and tribal chiefs, or on the mass of the peasantry, to rise in revolt; not only the *Limbus* but also the *Tamangs* did indeed rise up, largely for what might be termed political reasons, effectively refusing to accept the authority of the Gorkhali state. The Gorkhali state responded to the revolt of the *Tamangs* rapidly and brutally. Whelpton (2013, p.31) cites a royal order of 1793 issued with reference to the *Tamangs* of the 1,500 rivers region, in what are today Rasuwa and Nagarkot:

> When people belonging to the *Murmi (Tamang)* community engaged in rebellion *(kul)*[9], they were captured and beheaded. If any of the rebels is hiding there, seize him and hand him over to our *amali* (government official or judge). If it has been proved that

---

[8] See Terray (1974) for an African case study; also Lacoste (1974) for a case study in the Maghreb.

[9] The translation of *kul* here as 'rebellion' clearly makes sense, but it normally refers to a clan or family, or to 'all' or 'the totality', and so it is not entirely clear what was meant precisely.

you were not involved in the rebellion, no action will be taken against those who were not so involved (from Regmi, 1984,p. 129, quoted in Holmberg, 2006, p.36).

Whelpton comments that 'repression was the immediate response in Limbuwan as well, but the government in Kathmandu later adopted a more conciliatory policy, promising amnesty to those who had taken part in the rebellion if they returned to their former homes' (2013,p.31). The difference between the treatment of the *Tamang* rebels and that of the *Limbus* is striking. Whelpton comments, tellingly, as regards the latter, that 'the key problem was that with the low population density in Nepal at this time, it was difficult to replace cultivators who fled to another territory, and this meant loss of revenue for the central government' (p.31).

Indeed, one of the major policies of the Gorkhali state at the end of the 19th century, was to encourage immigration, land clearance, settlement and cultivation so as to maximize revenues from those occupying the land, for the pacification and administration of yet new territories, thereby establishing a positive feedback system leading to further expansion. This meant that, in reality, a combination of 'stick and carrot' was adopted for many decades, in order to create a viable and prosperous economy and society. Where the rural population was settled and within 'reach', the state was able to maintain effective control; further afield, and particularly where land was in plentiful supply and the local farming population sparse, immigrants who came as cultivators were generally welcomed and given concessions; only when established were they taxed.

The state was always, even in later periods under the Ranas, cautious about over-taxing the peasantry, particularly in sparsely populated and remote areas. The effective combination of the 'stick and the carrot' may explain the apparent relative lack of peasant resistance, unrest and revolt, throughout the 19th century, to which Jung Bahadur Rana referred.

The Consolidation of the Gorkhali Regime

Whelpton suggests that 'once the turmoil of 'unification' and its immediate aftermath was over, the mass of the population did, by and large remain in the state of passive obedience, which Jung Bahadur described to Orfeur Caveagh' (2013, p.38). He then suggests, vaguely, and without specific references, that 'there must have inevitably have been small-scale instances of rebellion at village level and at least in some areas, the possibility of a serious outburst does seem to have influenced government policy' (*sic*). He does not, however, identify any specific 'instances of rebellion at village level', or indeed refer to any other instance of peasant revolt until this uprising in Gorkha in the 1870s, led by a local *Magar* leader who apparently claimed to be a reincarnation of the saint Lakhan Thapa.

One explanation for this may simply be that the evidence of peasant unrest in the immediate 'post-unification' period has not been examined in sufficient detail; another may be that there were, in fact, relatively few instances of rural revolt or rebellion. If the latter is the case, then it may be related, as we have already argued, to the fact that, although during this early period the expanded Gorkhali state was concerned to increase the revenues accruing to the state, the sparseness of the population meant that the priority in agriculture was to encourage immigration, land clearance and cultivation, with a view to raising revenues largely by extending the area under cultivation rather than by raising the level of taxes.

Under such circumstances, when the extension of cultivation rather than increasing the level of taxation was the preferred option (and the dominant mode of production was in any case one of extensive cultivation combined with livestock production and the exploitation of 'wild' resources), the locals could often move out of range of state tax administration and collection. It is also possible that, particularly in areas where they were incomers rather than long-established ethnic groups, as was the case in some hill areas and in the *Tarai,* they were not politically organized and had only 'the weapons of the weak' with which to respond to over-burdensome taxation by the state – i.e running away rather than resisting. As the population increased, however, and settled farming began to become the dominant mode of production, this form of tax avoidance became increasingly difficult, the level of taxes increased and it seems that the exploitation of the peasantry increased.

The confiscation of *birta* land from *birta* holders, which began to gain momentum during the first decade of the 19th century and was implemented with a view to extending state control over newly reclaimed land for revenue purposes – often, it appears, resulted in resistance and rural unrest. The confiscation and conversion of *birta* and other land 'had a far-reaching impact on the status and condition of the cultivating class' – including the peasantry: 'such conversion did not necessarily lead to the eviction of existing occupants, but they were reduced to the status of tenants on the land which they had previously cultivated on free-hold tenure' (Regmi, 1999, p.75). Many former independent peasants found themselves obliged to pay rent, usually under the *adhiya* (half and half) system, on the land they cultivated, and their incomes tended to fall accordingly.

The response was often, initially, a complaint to the authorities; if and when this proved futile, it often resulted in the abandonment of farm and field, and emigration from the locality, so that some areas became de-populated; but 'the protest of the dispossessed *birta* owners to the 1805 confiscation measures did not long remain passive. Particularly in Tanahu and other areas in the western hill region, violent clashes broke out between them and the government officials deputed to administer the measures. The government understandably adopted a strong attitude towards such opposition. The ring-leaders were arrested and brought to Kathmandu in chains' (Regmi, 1999, p.77).

Officials sent to confiscate *birta* lands in Syangja district were directed to arrest and put in chains the *Brahmins* of Karkineta, who had promised to surrender their wives, their names, their lands and their lives all at the same time, and had gone into hiding. The officials were informed that if any *Brahmins* were to be killed in the course of this action, they (the officials) would not be held to account. Despite or arguably because of this, the protest organized by the *Brahmin birta* owners in the western hill regions became so acute in late 1805 that the government even banned their entry into Kathmandu without passports (Regmi, 1999, p.77).

There were further changes to the system of rural administration in the second decade of the 19th century. In 1812, the *adhiya* system was finally discarded as the basis for rent assessment in favour of the *kut* system (of fixed rents) in the Kathmandu Valley and large parts of the hill regions; in 1816, the system was further clarified. The level of *kut* rents was usually, however, higher than that of *adhiya*, and if output was lower than usual in a particular season, the peasant was still obliged to pay the same fixed rent. Furthermore, peasants could be evicted if they did not pay; although there is contradictory evidence in this respect (cf Regmi, 1999, pp. 89- 92). This created new difficulties, and contributed to a process of what might be termed class formation. *Kut* land grants were sometimes made to intermediaries rather than to actual cultivators, giving them effective jurisdiction over the peasantry and encouraging what Regmi significantly but arguably misleadingly terms 'sub-infeudation'.

Regmi states that 'the evidence is thus clear that the condition of the peasantry deteriorated to a considerable extent during the period after the political unification of Nepal' (1999,p.99), although it is likely that the extent to which this change occurred varied considerably from place to place. One consequence was an increase in the number of cultivators leaving their holdings; another was the occasional localized outbreak of peasant revolt. Regmi notes that, 'the government attempted to mitigate their difficulties by controlling the rates of payments collected from them and giving them a measure of tenurial security'; but, even accepting this rather benign interpretation of the measures adopted by government, it was also the case that 'rent and tenancy legislation intended to benefit the peasantry is always difficult to enforce, particularly in a situation where there is a close conformity of association and interest among revenue officials, *birta* owners and *jagidars* and the more powerful elements in the village'[10].

In addition to the burden of taxation and rent, peasants were also subject to demands for 'obligatory services' (*jhara*) of various kinds, although Regmi considers – significantly - that 'the Gorkhali rulers appear to have tried from the very beginning to insure that the growing burden of *jhara* services did not alienate the peasantry', and refers to several measures intended to mitigate or compensate for such demands (Regmi, 1999, pp.113-116). He notes, yet again, that such measures were applied mainly 'in areas which were sparsely populated or inhabited by unruly communities, and were situated close to the borders' – that is, where there was the greatest risk that demands for *jhara* services on top of everything else would result in peasants simply abandoning their fields and migrating elsewhere, or revolting, or threatening security.

He suggests, for example, that 'it can hardly be a coincidence that areas between the Koshi and the Tista Rivers (where compensation for *jhara* labour was provided) were inhabited by the *Limbu* and other *Kirati* communities who had not yet fully accepted subjugation by Gorkha (Regmi, 1999, p.116). So, it seems that rebellions did occur from time to time in different parts of the country against the Gorkhali regime in the immediate 'post-unification' period. In some instances, usually where the local population was settled and within the effective 'reach' of the state (as in the western hills), they were put down with considerable brutality. In his discussion of slavery in the post-unification Gorkhali state: 'rebels were lucky if they were enslaved, for capital punishment was usually awarded to all those who were above the age of 12 years'.

The 'Post-War' Period

Between 1820 and 1840, further experiments were made in the field of revenue administration, whose main objective was to ensure a stable basis for increasing revenue for the government. Firstly, the situation in the

---

[10] Regmi himself admits that 'at the same time, one would suspect that the measures undertaken by the government from time to time 'for the benefit of the peasantry' were not always meant to be actually enforced' (Regmi, 1999, pp. 99-100).

Tarai had to be regularized and overhauled, and the border between India and Nepal demarcated and agreed; secondly, the situation in the rest of the country had to be reformed. The strategy involved the development of a more permanent structure for local administration and tax collection. In the hill regions, including the Kathmandu Valley, experiments undertaken between 1820 and 1827 included both arrangements ('settlements') with individual *ryots* and also with village headmen for the entire village as a unit. In 1828, the appointment of local revenue functionaries on a five year basis was tried in the eastern *Tarai* and then continued in 1834 for another five year term. In 1838, revenue functionaries in the western *Tarai* were appointed on the same basis. All of this suggests a process of uneven development in the administration of the Asiatic mode of production in the Gorkhali state.

The general trend as a result of all this was an upwards movement in rents, which tended to disadvantage the peasantry, combined with a gradual shift from in-kind payments to payments in cash, which suited all parties, but forced peasants increasingly to sell at least some of their output to meet their fiscal obligations even as the government was attempting to fix the price of rice and other agricultural commodities, 'mainly with the objective of lowering the costs of feeding the army' and, more generally, maintaining the growing bureaucracy (Regmi, 1999, pp.179-185). The level of rents determined in the course of the 1837 rent settlement appears to have been the highest ever in the history of Nepal – in the hills and in the *Tarai* (Regmi, 1999, p.185).

Those who were unable to meet the tax requirements were in trouble. Some were evicted; others chose to abandon their fields and emigrate[11]. Regmi suggests that 'the post-war period in fact, witnessed a large-scale exodus of people from several parts of Nepal to Indian territory' (Regmi, 1999, pp.194). Local people also complained and, in some cases apparently, revolted. During the 1830s, the government was even compelled in some instances to re-think its policy; but overall it stuck to its guns - literally.

---

[11] Regmi notes that '*Pallokirat* was one of the areas worst affected by the problems of indebtedness and emigration' (Regmi, 1999, p.194, note 163).

Payment was often obtained by the use of force and, if necessary the army was deployed.

On the other hand, those who managed to make their payments and maintain cultivation were rewarded with more secure land occupancy rights (Regmi, 1999, pp.185-86). Also, *jhara* (forced labour) for various services declined and there was a trend towards payment for portering and other labour previously provided by locals 'for free'. At the same time, during the late 1820s and the 1830s, measures were taken to improve the condition of slaves and bondsmen, in part to attract back those who had fled punitive taxation and rents, or escaped from slavery and/or bondage to India, to return and reclaim waste land for settlement (Regmi, 1999, pp.189-191). This suggests an early stage in the progressive transformation of the Asiatic mode of production into something different, which would be increasingly based on the reality of 'private' land ownership.

It was still government policy that land was to be reclaimed and arable land used primarily for subsistence by peasant farmers. Still, the government did not officially permit rent-receiving land ownership, except through *birta* grants or *jagir* assignments. This, it could be argued, suggests the continuation of the Asiatic mode of production. In reality, however, the situation at the local level was complex, and the simplistic notion of 'the state' directly confronting and exploiting 'the peasantry' somewhat misleading. Throughout this period, and indeed even in the previous period of 'unification' to some extent, cultivators could be owners, tenants or sharecroppers, and there was a plethora of 'intermediaries', some locals and some officials, between 'the state' and 'the peasantry'. Local intermediaries included those *chuni ryots* who possessed such extensive landholdings with jurisdiction over the lands occupied by others, including often entire villages, that they were called *zamindars* (Regmi, 1999, p.32 note 80).

This was a category of state official originally introduced in 1793 to reform tenurial conditions in both the *Tarai* and the hill regions, following reforms undertaken in early that year by the British East India Company. There were large numbers of *zamindars* in both the eastern and western *Tarai,* and also in some parts of the Karnali region; they paid land taxes to the state but their land was cultivated by sharecroppers (Regmi, 1999, pp.32-33). Even so, Regmi emphasizes, importantly, that 'while *zamindars* in India became proprietors of the land, their counterparts in Nepal remained nothing more than revenue collection functionaries' (Regmi, 1999, pp.93). By this time, however, even if formally they were not private landowners, they were, in effect landlords, extracting surplus from the peasantry under their jurisdiction in the form of rent, and paying taxes to the state.

In the far western hills of the Karnali region, where land taxes were generally paid in cash, *jagirdars* were initially appointed and paid salaries in cash to oversee the collection of taxes by the commander of the local military unit; but from 1805 onwards, lands were actually assigned to them as *jagir*, with jurisdiction to collect taxes/rents directly themselves from the peasantry, or (in some cases) from a 'sub-tenant' – leading to what Regmi (significantly) calls 'sub-infeudation'[12].

In the central and eastern hills, by contrast, this level of authority was matched only by the headmen of *kipat*-owning Kirat communities, which resembled those formations described variously by Marx in the *Grundrisse* as 'tribal' and 'Germanic'. In the western hills also, it is possible that, despite the prevalence of peasant communities and virtual absence of *zamindars*, village headmen exercised a somewhat similar, though probably more constrained, authority over local villagers. Indeed, Regmi discusses the range of 'levies' collected by village headmen in the hill regions for various purposes (Regmi, 1999, p.66).

Some of the indigenous (*adhivasi*) modes of production remained relatively 'primitive' in terms of their technology, largely involving fishing, hunting and gathering; others were more developed and included shifting cultivation and livestock production; yet others were involved in more or less settled cereal agriculture. Some remained, in socio-political terms, at

---

[12] (Peasants cultivating *jagir* lands under conditions of 'sub-infeudation' occupied more or less the same position as those cultivating *birta* and *guthi* lands); Regmi remarks that 'they rarely find a place in the revenue literature of this period, except in so far as they were considered a part of the general mass of the peasantry' (1999, p.97).

the level of 'communities' or 'bands'; others had a more complex structure and were organized in clans and/or tribal groups, sometimes with village elders and/or clan heads; yet others were more hierarchical and had tribal chiefs and rulers, and identified themselves as 'ethnic groups'. It is clear, therefore, that at the local level there were still various different 'modes of production' in operation, which, while embedded within a wider political economy, nevertheless had their own dynamic, and also had implications for the operation and development of that wider political economy.

Whatever their mode of production, and their social and political characteristics, in so far as they were for the most part cultivators subordinate to the Gorkhali state, most of the population could be broadly characterized as peasants (in Wolf's terms), subject to state jurisdiction and authority, and obliged to pay rent and taxes. The Gorkhali state had already, by the end of the first few decades of the 19th century, evolved into a powerful apparatus, presiding over an assemblage of social formations, characterized by its overall capacity to control and coerce, if needs be, and to raise revenues. This was not a feudal state, but rather what Marx referred to (inadequately) as an 'Oriental state' with an 'Asiatic mode of production'. This was to be consolidated under the Ranas, progressively, into a distinctive form of 'prebendalism'[13]

Rana Rule

Regmi comments that:

> 'The emergence of Rana rule marks a turning point in the history of Nepal. To be sure, this event *per se* did not herald an era of change. It would not even be correct to say that the Rana's economic policies and programmes marked a radical departure from those followed by their predecessors. Rather, that Rana take-over of power synchronized with a number of internal and external developments which made it possible for

---

[13] Prebendalism refers to political systems where state officials (civil and military) feel they have a right to a share of government revenues, and use them to benefit their own extended families and social networks. See Gallissot for a discussion in relation to pre-colonial Algeria.

the Rana rulers to adjust their economic policies and programmes in a more effective manner than their predecessors' (Regmi, 1988, p.3).

The external developments could be summarized as the settlement achieved with the British in the three decades after the Anglo-Nepali war (1814-1816), whereby the territory of we can now call the state of Nepal was more clearly defined and delimited, and the Ranas were assured – and took measures themselves to assure – that their rule over this territory would not be contested, at least by the British; but were obliged to accept a British representative (a Resident) in the capital. Nepal became 'a semi-colonial state' – a term used by <u>Lenin</u> and Mao Tse Tung to describe states that in the late nineteenth and early twentieth centuries were penetrated by imperial capital, trade, and political influence, but which preserved their juridical and formal political independence[14].

As regards internal developments, Regmi suggests that:

'The (domestic) history of Rana rule in Nepal is basically an account of how an impoverished family, which nevertheless ranked high in the social and political scale, and belonged to Gorkha's traditional military aristocracy, was able to grab political power and use that power to build up a centralized state and administration' (1988,p.5). He argues that 'the government, under the Rana political system, merely functioned as an instrument to carry out the personal wishes and interests of the ruling Rana Prime Minister, and its main domestic pre-occupation was the exploitation of the country's resources in order to enhance the personal wealth of the Rana ruler and his family' (1988, p.6).

He explicitly contrasts this with that prevailing in pre-industrial England, where landed property was the foundation of political authority and where, even at the beginning of the 19th century, Britain was still ruled by landowners (1988, p.6). The term 'feudalism' is, therefore inappropriate, as a characterization of the political economy of Nepal under the Ranas, at least as far as Regmi is concerned. On the other hand, he suggests that

---

[14] Other examples include Persia, China, Thailand, Afghanistan, Yemen, and Ethiopia.

'the functions of both revenue collection and civil administration largely belonged to the landowning elites' (1988, p.19)[15]. The development of the administration of rural areas and of the army were both features of the early Rana period, leading the consolidation of control over the rural areas by a bureaucratic state, serving the interests of a ruling elite – a form of prebendalism, or what some have called 'command feudalism'[16].

The 19th century Nepali state paid emoluments to its employees and functionaries partly or wholly through assignments of taxable lands and villages rather than through payments of money; such assignments were known as *jagir*. The system was used on such a wide scale that legislation banning the payment of cash salaries to government employees so long as taxable lands were available for assignment as *jagir* remained on the statute book throughout the period of Rana rule, albeit with varying degrees of effectiveness (Regmi, 1988,pp.29-30). In addition, there were four main categories of what Regmi calls 'intra-state agencies' responsible for revenue collection: *rajya, birta, guthi* and *kipat*.

Rajas of vassal principalities were also entitled to collect and appropriate revenues from the territories under their jurisdiction; these included Bajura, Bajhang and Jajarkot in the far-west and Mustang in the north-west. *Guthis* were grants made to monasteries and temples, or to other religious or charitable associations. *Kipat* was a system confined to the *Kirat* ethnic groups, notably the *Limbus* of the far eastern hills. The *birta* system was different; under this system, arable land, forest, and villages were granted to individuals on a tax-free basis, enabling them to

---

[15] I would prefer to refer to these state functionaries as 'landlords', rather than as 'land owners' (which, by and large, they were not) as their ability to extract surplus from the local peasantry in the form of rent was by virtue of their appointment and/or recognition by the state.

[16] Gallissot uses these terms to describe pre-colonial Algeria. He argues that 'although the fiscal character of exploitation on the one hand, and the public and military nature of the social category at the head of the hierarchy on the other, give as Asiatic aspect to the society', the level of development of the productive forces and the class structure suggest that 'pre-colonial Algerian society is truly feudal, distinguished only by one singularity: the primarily fiscal character of the oppression' (Gallissot, 1975,p.438).

appropriate surpluses directly themselves; this system, then, allowed for the development of land 'owners', as well as landlords.

Less than three months after Jung Bahadur Rana became prime minister, he made arrangements to compensate former *birta* holders for the *birta* land confiscated in 1805-06. But few, apart from the Ranas themselves, were able to take advantage of this arrangement - and it was not until 1882 that Ranoddip Simha Rana took effective measures to compensate other *birta* holders through allotments of taxable *jagera* land. *Jagirs* absorbed a sizeable proportion of revenues, but the Ranas appropriated the lion's share of *jagir* incomes to themselves; and the largest beneficiary was the prime minister himself. This is a classic characteristic of 'prebendalism'.

In addition to these various mechanisms for raising state levies, were Crown levies, which accrued specifically to the royal household and the palace. The monarchy was retained, but was marginal in most respects to the political economy of Nepal, serving a largely legitimizing and religious role. Unlike the sultanates and sheikhdoms of the Gulf, Nepal under the Ranas was ruled not by the Shah royal family but by a family dynasty of hereditary prime ministers.

Suggestions that the political economy of Nepal under this regime was 'stagnant' in any way are misplaced and mistaken. This was no static and unchanging Asiatic mode of production of the kind identified by Marx and Lenin; it was an authoritarian regime with a state apparatus (army and administration) committed to increasing revenues for the ruling class as a whole, and particularly for the ever-expanding Rana family, associated with a growing and relatively dynamic agrarian economy which benefited from its relationship with British India by exporting agricultural and forest-derived raw materials and importing manufactured and luxury goods on an increasing scale.

State revenues rose between 1851 and 1861 from around NRs 1.4 million to NRs 3.5 million or two and a half times more than previously; in 1877, the last year of Jung Bahadur's regime, it was estimated by Daniel Wright that the total revenue of Nepal was about NRs 9.6 million. Nearly ten years later, G H D Gimlette felt that NRs 10 million was 'nearer the mark'. Even taking inflation into consideration, this was a substantial increase in real terms. One explanation was that government policies and incentives to promote immigration, land clearance, settlement and cultivation were by and large successful, and the area under cultivation available for taxation increased. A second is that the efficiency in the process of administration and tax collection increased significantly as the Rana regime became increasingly bureaucratic and the army was deployed effectively to maintain overall control. A third is that revenues were increased as a result of the export of timber and other forest products on an increasing scale as demand grew in India.

In fact, the evidence suggests that all three of these provide elements of an explanation. As far as the first is concerned, Regmi remarks that 'thanks largely to immigration from India, the cultivated area in the *Tarai* region increased steeply during the half-century of Rana rule. The trend inevitably had a positive impact on the role of that region as the major source of state revenues' (1988, p.114). As for the second, Regmi (1988, pp.60-90) provides ample evidence that the early Ranas built the foundations for an effective system of fiscal administration, particularly in the *Tarai*, that lasted 'for well over a century until the 1960s' (1988, p.77).

This included the appointment of *jimidars* (a new category of revenue functionary) for each *mauja* (the basic administrative unit below the district) and an increase in the number of army battalions responsible for supporting the fiscal administration. Regmi concludes that the system of *jimidars* operating at the level of the *mauja* and the 'the establishment of a permanent revenue collection office at the district level was the most important Rana contribution' (1988, p.90) to fiscal administration. The bulk of revenues so collected were from the *Tarai*, and particularly from the eastern *Tarai*, which contributed nearly three out of every four rupees (Regmi, 1988, p.93).

A fourth factor in the equation is that it seems that, at least in some instances and in some areas, the level of taxes on the peasantry increased in real terms. The response of the peasantry to the tax burden imposed upon them was manifold, but does not appear to have often involved outright revolt. At first, it appears, they tended to appeal against increases, when and where they were introduced. Often, it seems, the state was willing to heed their concerns, and respond to them. Regmi insists that 'the Ranas appear to have followed a very circumspect policy in the field of agrarian taxation. Their primary concern was not to increase the volume of agrarian tax revenues but to ensure that it did not fall below current levels' (1988, pp.118-119); and in quite a few cases there is evidence that the regime was open to the concerns of the local peasantry as regards tax increases.

For example, Bir Shumshere appears to have raised agrarian tax rates in the western *Tarai* region only once during his fifteen year rule; the increase, approved in 1892, amounted to 2 paisa in the *rupee*, or just over 3 per cent' (Regmi, 1988, p.119). It is my view that the regime was concerned to increase total revenues; but wherever possible by the old policy of promoting the extension of cultivation, mainly in the *Tarai* and mainly by importing capital and labour from India – in other words, promoting development by an investment in capital, land and labour (as Dahal has argued), rather than be squeezing the peasantry. Indeed, Regmi effectively accepts this argument in his concluding discussion (Regmi, 1988, p.263).

On the other hand, it remained the case that when tax increases became too onerous, particularly where there was little scope for further extension of the area under cultivation, the response by the peasantry was emigration. The evidence strongly suggests that, towards the end of the 19th century, as the limits to the extension of cultivation were reached, and population density began to increase, particularly in the hill regions, migration became an increasingly common response to growing taxation and indebtedness, as it was to be during the 20th century.

There is little evidence of peasant unrest in the *Tarai* during the Rana period, largely it seems for these two reasons. Firstly, the state was cautious in increasing taxes to the point where there was a risk of rural unrest; and secondly, perhaps, because of the 'safety-valve' provided by an open frontier with British India to the south. There were, however, some signs of dissidence in both the western and the eastern hills; these, however, were more in the form of elite politics and local rebellion against the regime than in peasant unrest.

The Revolt of Lakhan Thapa Magar

The revolt of Lakhan Thapa Magar has been examined in some detail by Marie Lecomte-Tilouine (2000, 2003). Whelpton confines himself to a terse passing comment: 'this was a messianic revolt in which a subordinated group saw a chance of turning the tables on its overlords with divine assistance, but it did not take hold on a large scale and his (Lakhan's) house was swiftly surrounded by government troops, and Lakhan Thapa himself was hanged' (2013, p.39). There is quite a lot more to this revolt, however, than that short summary suggests.

In 1874, Lakhan Thapa Magar began to mobilize opposition to the regime of Jung Bahadur and in 1876 launched a protest movement from Bunkot (Bungkot), Kahule Bhangar in Gorkha. It had a degree of popular legitimacy, and was able to mobilise the peasantry at least to some extent. A poem by Gyan Dildas, who supported the revolt, spoke out against the way in which the regime had strengthened the position of the rich and failed to provide justice (*adam bajyo dhaniko firyo jaga janha, ghusyaha bichari nishaf herchha kaha*). Lakhan Thapa also claimed legitimacy as a local leader by claiming to be the reincarnation of a local saint, according to some accounts and of the Goddess Manakamana Mai of Gorkha, according to others.

Pramode Shamshere Rana suggests that 'his preaching attracted a sizeable following for his mission to overthrow the Rana rule. About two thousand young men rallied behind him and he managed also to get (hold) of some small arms' (Rana, 2000,p. 49). As the movement grew, Jang Bahadur Rana came to know of it and sent the Debi Dutt regiment to Gorkha to suppress the rebellion. Most of the rebels were rounded up after a brief skirmish; more than 300 rebels – the majority of them probably local peasants mobilised by Lakhan Thapa - were apparently killed in action, but many others surrendered. Accounts suggest that Thapa and a handful of his colleagues were captured during a secret meeting; some 60 guns and ammunition were also apparently found and seized. The rebel leaders were put in bamboo cages and brought to Thapathali Darbar, where a summary trial was held. They were whipped and forced to sign confessions. They were then hanged in front of the Manakamana Mai temple back in Gorkha on 27 February 1877.

It is hard to know how best to characterize this movement. It could be seen as a part of the elite opposition that Whelpton identifies - given undoubted links between events that took place at the time of the *Kot* Massacre, when several high-ranking *Magar* army officers were killed either directly or indirectly by Jung Bahadur Rana, and this revolt (although admittedly it did take place nearly 30 years after the *Kot* Massacre) - or as a local rebellion. If the touchstone of a peasant revolt is that it is rooted in the land question, then this was certainly not a peasant revolt; although Pramode Shamshere Rana suggests that 'it was the first public revolt against the dictatorial rule of Maharaj Jung Bahadur: it was a revolt by the people' (2000,p.50). It certainly drew a degree of legitimacy from the accusation that 'the regime had strengthened the position of the rich, and failed to provide justice', and from Thapa's claim to be the reincarnation of a local saint, but it was not, I suggest, a classic peasant revolt.

As Whelpton remarks, '*Magar* activists have portrayed Lakhan Thapa as a proto-democratic fighter against the Rana tyranny and, in 2000, the Nepalese government officially declared him a 'martyr'. The religious dimension of his revolt, however, suggests that he had more in common with earlier traditions of resistance to central authority than with twentieth-century political activism' (Whelpton, 2013, p.69, footnote 20). I am not myself convinced that the so-called 'religious dimension' was much more than a device to give the opposition more legitimacy among the local population. There are numerous instances around the world of rebellions led by men who claimed to have divine support for their action, or in some cases claimed themselves to be religious figures, prophets or saints[17]. The movement can be more appropriately characterised, I suggest, as a rural rebellion, led by local notables[18].

Jung Bahadur Rana himself died the same year, in 1877, bringing to an end a distinctive period in Nepal's history. There followed nearly a decade of plot and counterplot to determine the succession, pitting brothers against each other and cousins against cousins. Finally, in 1885, the closest male relatives of Jung Bahadur Rana - his brother Ranoddip, his son Jagat Jang and his grandson (Jagat Jang's son), Yuddha Pratap - were assassinated by the Shumsheres, the sons of Jung Bahadur's brother Dhir Shumshere, who then ruled Nepal for the next 65 years. Jung Bahadur's remaining close relatives first took refuge in the British Residency, and then managed to secure safe passage to India. Later, they attempted to launch an armed attack against the new rulers; but this was unsuccessful and several of them were then put under supervision by the British Indian authorities. This was perhaps the clearest example of 'elite politics' since the *Kot* massacre nearly half a century before.

As regards more broad-based rural unrest in the latter part of the 19th century, Whelpton notes that 'the *Limbus*, a much more homogeneous group than the *Rais* to their west, succeeded better than other *Tibeto-Burman* groups in retaining their traditional *kipat* system of communal landholding, at least up until 1886'. In that year, however, legislation was

---

[17] See, for example, Bu Hmara, the 'man on a donkey' who led a rebellion against the Sultan in pre-colonial 19th century Morocco and claimed to be the legitimate Sultan and to have supernatural powers (Seddon, 1978).

[18] More like Kett's rebellion: a revolt in Norfolk, England during the reign of Edward VI, largely in response to the enclosure of land. It began at Wymondham on 8 July 1549 with a group of rebels destroying fences that had been put up by wealthy landowners. One of their targets was yeoman farmer Robert Kett who, instead of resisting the rebels, agreed to their demands and offered to lead them. Kett and his forces, joined by recruits from Norwich and the surrounding countryside and numbering some 16,000, set up camp on Mousehold Heath to the north-east of the city on 12 July. The rebels stormed Norwich on 29 July and took the city. But on 1 August, the rebels were defeated by an army led by the Marquess of Northampton, who had been sent by the government to suppress the uprising. Kett's rebellion ended on 27 August. Kett was captured, held in the Tower of London, tried for treason, and hanged from the walls of Norwich Castle on 7 December, 1549.

passed which provided for the conversion of all land granted by *Limbus* to non-*Limbus*. There was, as a result of this government intervention, 'rising *Limbu* discontent' over the amount of land now being permanently lost to 'their community' – and in some parts of *Limbuwan* such was the level of dissension and dissent that it proved impossible to complete revenue surveys after 1893. The effective refusal of the government's proposed legislation was a form of rural protest, resistance verging on rebellion. Caplan states that there followed several attempts to organize opposition among the *Limbus* to stem the tide of *kipat* conversion to *raikar* tenure by the government (Caplan, 2000, p.175, cited in Karki, 2002, p.9).

The result, interestingly, was a major concession: new legislation in 1901 forbade the permanent alienation of any *kipat* land to a non-*Limbu*. But this must have proved ineffective, for in 1913 and in 1917, the *subbas* of Ilam forged an alliance to combat continuing land alienation. According to Caplan, over 300 men - mainly *subbas* – came forward and challenged the government's policies with regard to *kipat* tenure, leading to the withdrawal of proposals to abolish it altogether. The *subbas* were able to maintain this alliance, and in fact, formed an *ad hoc* committee to defend their traditional *kipat* tenure, throughout the first half of the 20th century. This is clearly another example (like that of the *birta* owners in the *Tarai* referred to previously) of local elites challenging the authority of the state, on behalf of their peasant communities, and of rural folk more generally defying the state.

Whelpton insists on framing this intriguing example of resistance to government land and taxation policy solely in 'ethnic' terms, which is, of course, how the leaders of the unrest characterised it; but it can also be seen as a peasant movement, in so far as it mobilized cultivators around the issue of land and land rights. This peasant movement was clearly associated with strong resistance by local leaders and peasant farmers, with their own distinctive form of *kipat* communal tenure, to government intervention to legitimise the alienation of their communal heritage.

The ability to maintain effective organized local resistance to intervention of this kind by the Rana regime was clearly linked to the communal basis of land tenure in the far-eastern hills, the relatively egalitarian nature of local economy and society, and the tradition of local chiefs and resistance from the time of 'unification' onwards to further integration into the state. In other words, their ability to resist further encroachment was linked to a combination of political ideology (ethnicity and egalitarianism) and political economy (community-based economy and society). It was also, on the other side, linked to the difficulties of deploying state force to ensure compliance in such a difficult and relatively remote region against determined local resistance. In other parts of the Rana state, the 'reach' of the state was greater and more effective.

The eastern hills were to see yet another instance of rural protest when, in the early 20th century, a middle-aged Brahmin widow from Bhojpur led a movement, opposing caste discrimination and oppression, and against the many other forms of state injustice, which lasted from the second decade of the century until the 1940s. This is one of the few rural protest movements led by a woman and with large numbers of active female participants, although women have always been important, if often invisible, participants in rural protest[19].

Yogmaya Neupane was born into a *Brahmin* family in Nepaledada VDC in 1860, and married very young. She was widowed within three years of her wedding, badly treated (as a *poi-tokuwi*) by her in-laws, and returned home. She developed a secret liaison with a local boy and eventually left home (and Nepal) to marry him in Assam. When he died some ten years later, she married again. Widowed for a third time, Yogmaya, who by now had a daughter, Nainakala, decided to adopt a life of renunciation and live the rest of her life as an ascetic. While it was fairly common for males to become ascetics in Hindu society at the time, it was very rare for females to do so.

She returned to Nepal with her daughter to her home village, and handed over daughter Nainakala to her brother and his wife. In this way, she gave up all her responsibilities and assumed the life of an ascetic. It was during this period that Yogmaya began composing her religious

---

[19] This account relies in part on accounts published by Barbara Aziz (1993, 2001).

songs and poems[20]. Analysis of her early works suggests that she was heavily influenced by the principles of reformist Hindu leader Dayananda Saraswati, who was very popular in India at the time Yogmaya was living in Assam. During 1917 to 1918, she travelled as an ascetic within Nepal and came in contact with many renowned religious leaders, including Swargadwari Mahaprabhu Abhayananda Second, who admired her devotion and guided her in her Yogic spiritual education. She then returned to her home village to practice her *sadhana.*

Her previous personal experiences, and the discrimination she had faced as a widow, deeply affected her outlook on what she saw as the injustices of Hindu society in Nepal. Despite being a follower and preacher of Hindu spirituality, she made it clear in her songs and poems that the patriarchal caste-based society was unacceptably harsh towards women, lower caste groups and the poor and disadvantaged more generally. Her literary works, which were directed at the women, peasants and workers of Bhojpur, are well reflected in the poem, '*Maharaj Chhan Darbarma Herna Aaundainan; Dukhi Janle Niya Nishaf Sidha Paudainan*', which criticizes the regime for its lack of compassion for the suffering of the poor and lack of justice. Her particular focus was the exploitation and oppression of women.

This resonated with an increasing number of people – mainly women. She formed and became the first chair-person of the Nepal *nari samiti* in 1918. Her local following grew through the 1920s, and after the publication of her poems, people from as far away as Darjeeling and Kathmandu began to join become her disciples; she now had thousands of followers. The local elite – including landowners and state officials - began to find her a significant threat and attempted to contain her influence. In 1931, Yogmaya sent one of her disciples, Prem Narayan Bhandari, to Kathmandu to spread the word. Learning of this, the then Prime Minister of Nepal,

---

[20] Yogmaya and her daughter Nainakala were illiterate. Her poems were compiled and published by her literate disciples. In fact, the *Yogbani* is considered to be only a small part of her collection of poems, which she continued to compose until the year of her *jal samadhi* (1941).

<u>Juddha Shumshere Rana</u>, met with Bhandari and promised he would take Yogmaya's message seriously.

There was little change, however, over the next four years and so, in 1936, Yogmaya once again sent Bhandari to Kathmandu; and this time she herself also travelled to Kathmandu, with her daughter. During her visit, she had an audience with Juddha Shumshere, at which he reportedly asked what she wanted, and she replied: 'the alms of the holy order of truth and justice'. Then, before leaving the Kathmandu Valley, Yogmaya handed him a 24-point appeal detailing the reforms she and her followers wanted to see in the country. This alarmed Juddha Shumshere and his circle, who, already concerned by growing political dissent, began to see Yogmaya and her movement as an additional threat.

Already, the regime had grounds for concern: increasing discontent among the Nepali intelligentsia at home and abroad from the 1920s onwards, the establishment in 1932 of the *Prachanda Gorkha* – a group of young Nepalis who planned to bomb all senior Ranas, but were arrested before they could implement their plans, and then exiled or jailed - and the formation in 1935 of the *Praja Parishad* – a party that espoused democracy and a constitutional monarchy. Back in Bhojpur, Yogmaya and her movement were now constantly monitored.

For her part, Yogmaya was becoming increasingly vocal in opposition to the regime, declaring that now it was time to begin a new era by destroying the injustices, superstitions and corrupt practices that so adversely affected Nepali society. She added that she herself was nearing her own 'salvation' and would demonstrate this by preparing for a human sacrifice in which she and hundreds of her disciples would be burned to death in a ritual called *agni samadhi* on Kartik Shukla Purnima (12 November 1938). Fearing widespread rural unrest if such a mass sacrifice were allowed to happen, Juddha Shumsere ordered the disruption of the event by deploying around 500 security personnel at the selected place on 11 November, following which eleven male disciples were jailed at <u>Dhankuta</u> and most of the female would-be participants in the *agni samadhi* ritual, including Yogmaya and Nainakala, were confined to the Radhakrishna Temple at Bhojpur.

The women were released after three months, and most of the men three years later in April, 1941. After the release of all her disciples, Yogmaya decided to continue with her self-sacrificing plans, but this time the plans were to remain a secret. She set a new date for a *jal samadhi* – 5 July 1941, which was the day of the holy Harisayani Ekadasi - and allowed only a select group of disciples to join her. The religious ritual for the *jal samadhi* began on the night of 4 July 1941 and in the morning of 5 July at around 4 am, Yogmaya led the ritualistic mass suicide by climbing on a rock placing a plate with lighted oil lamps on her head on the bank of the Arun River. After Yogmaya, some 65 of her disciples jumped on the river in succession that morning, and the next day two more followed suit. Ultimately, the total death count from the event stood at 68[21].

The Rising Tide of Political Opposition

In October 1940, the leaders of the *Praja Parishad* were arrested and charged with plotting the assassination of the Rana family. Four were sentenced to death and others to long jail sentences. This was, however, essentially a movement of disaffected sections of the emerging 'urban middle classes' and not based in rural discontent. It was not until the foment of the 'Quit India movement' in India to the south and the rise of

---

[21] Information about Yogmaya and the mass suicide was strictly censored by the Rana regime. Even after its overthrow, during the Panchayat era, it remained difficult to discuss her legacy. Yet, her disciples, mainly in the districts of Bhojpur, Khotang and Sankhuwasabha, maintained the tradition locally. Some of Yogmaya's female disciples continued to live in the Manakamana Temple in Tumlingtar. More recently, however, scholars, including Barbara Nimri Aziz, Michael Hutt and Dipesh Neupane, have carried out research and published their findings on Yogmaya and her rural movement. Topics on Yogmaya and her movement have been also included in the curriculum in specific Social Science based disciplines in different universities in Nepal including Sociology, Anthropology and Women's Studies. A local organization has been formed in Bhojpur, called the 'Yogamaya Shakti Pith Tapobhumi Bikash Tatha Vikas Sanstha', to promote her work and activities as well as preserve the places where Yogmaya spent a significant time during her ascetic life around Nepaledada and Dingla. On November 16, 2016, the Government issued a postal stamp in recognition of her contributions to the history of Nepal.

Indian nationalism spurred on the establishment of both the Nepali Congress and the Communist Party in India in the 1940s that a systematic opposition – in the rural as well as the urban areas - to continued Rana rule was created. Interestingly, the very first two lines of the Communist Party manifesto were: '*pahad parbat khola nala jamin sabkho; malik hami dash banau hisha shabko* (the mountains, rivers, streams and land are ours; we are the masters – why should we be slaves – we all have a share (in our heritage) (CPN, 1949, cited in Karki, 2002, p.9).

The communist movement of Nepal really began to take shape after the publication of the Nepali edition of the Communist Manifesto in April 1949 – the year that the Communist Party of China came to power and Mao Tse Tung began to re-formulate Marxist-Leninist thought to identify the peasantry, and especially the 'middle peasantry', as a revolutionary class. The Communist Party of Nepal (CPN) was formally established in Calcutta in September 1949. The objective of the CPN was 'to oppose the autocratic <u>Rana regime</u>, <u>feudalism</u> and <u>imperialism</u>'. The first leaflet produced by the CPN in April 1949 declared that Nepal should establish a 'new democracy', as in China – if necessary through armed struggle – so as to create a People's Republic'.

The next year, in April 1950, a merger of the Nepali National Congress (founded in 1947) under B P Koirala and the Nepali Democratic Congress (formed in 1949), led to the establishment of the Nepali Congress Party. Even before this - in 1947, according to Sadhana Adhikari (wife of Man Mohan Adhikari, quoted in Pandey & Rimal nd,p.14) - the Nepal Women's Association had been founded, and over the next two years demanded votes for women (which were won in 1952) and education for women. Apparently, large numbers of peasant women joined the Association, and in 1949 a peasants' conference was held in the house of her *nati dai* with more than 20 peasant women actively participating. Peasant women continued to be actively involved in the Nepal Women's Association throughout the next decade[22].

---

[22] The Nepal Women's Association appears to be different from the All Nepal Women's Association set up by Punya Prabha Dunghana in 1951.

In 1950, King Tribhuvan sought refuge in the Indian Embassy, and then fled to India. The rebel forces of the Nepali Congress crossed the border into Nepal and launched an attack on the district headquarters at Birgunj, where the garrison soon surrendered. Leaflets, and the party manifesto, written by B.P. Koirala in 1950, were dropped from the air over Kathmandu at the beginning of the revolt. These promised loyalty to a constitutional monarchy, the introduction of democracy 'as in the west' and social reforms. Throughout the next month, the armed insurrection of the *Jana Mukti Sena* (People's Liberation Army) developed and spread, picking up support from within the country, in part from the defection of Rana troops to the rebels.

The defection of the Palpa garrison in western Nepal in early January 1951 was a notable 'coup' for the rebels; but they also proved able to overcome or drive out the small and generally ill-trained army garrisons from the hill towns where they tended to be located. The rebels were supported by a local uprising against the Rana regime in the eastern hills. Civil unrest in Kathmandu, Pokhara and several towns in the *Tarai* also supported and complemented the armed struggle against the regime. In the meanwhile, the king and the Indian government came to terms on their own, ensuring that the overthrow of the old regime would not be as complete as the opposition had intended.

What came to be known as 'the Delhi Compromise' led to an agreement (between the Rana government, the King and the Indian government) whereby the king would play the role of a constitutional monarch, under a new democratic Constitution (to be framed by a Constituent Assembly elected by the people). The last part of this agreement was never honoured: an interim government was formed in February 1951 to carry out the task, but the king refused to allow the framing of a new democratic constitution, and his successor, King Mahendra, who came to the throne in 1955, also failed to honour the Royal Proclamation.

PART TWO

Peasant Movements in the 1950s

The early 1950s constituted a period of generalized dissidence characterized not only by the efforts of political activists from outside Nepal – the Nepali Congress and the Communist Party, for example – at the very beginning of the decade to remove the Ranas from power, but also by a plethora of smaller and less orchestrated peasant movements, some of them linked to these new revolutionary political parties and to the newly formed Peasant Associations – the All Nepal Peasant Association (ANPA), the Agriculture Reform Association (AFA) and the Worker-Peasant Association (WPA) - others not.

Suresh Thapa refers to the first two decades after the overthrow of the Ranas as 'the golden age of peasant movements in Nepal' (1996, p.75), involving struggles against landlords and also against the government. Pandey and Rimal also comment on the importance of peasant movements through the years from 1952 to 1958 but add that 'in these movements, mentionable is the action against the autocratic Bada Hakim Santbir Lama of Ilam. A woman activist Dulalni Bajei pulled Santbir off his horse and garlanded him with shoes. Many brave women like Dulalni Bajei are, however, not mentioned in history, their deeds have not come to public attention' (Pandey & Rimal nd, p.16). The extent and nature of the involvement of women in peasant movements and more generally in rural unrest certainly remain largely invisible, and much is to be done as far as research into this issue is involved.

The Eastern Hills

The armed insurrection or rebellion against the Ranas in the eastern hills at the beginning of the decade was largely organized by Rai and Limbu ex-servicemen, many of whom were also reasonably comfortable 'middle peasants', and was centred in Bhojpur, where rebels under Naradmuni Thulung (Rai) seized control on 9 December 1950, setting up a provisional government and dispatching forces to seize other towns in the eastern hills. A government force sent to re-capture Bhojpur was defeated and the commander summarily executed. The rebels extended their control eastwards into Sakhuwasabha, Dankhuta, Terathum and Taplejung; south through into the eastern *Tarai;* and westwards to Dhulikel on the edge of the Kathmandu Valley (Whelpton provides a map on p. 41).

Although there was undoubtedly communication between the Nepali Congress and Naradmuni Thulung's revolt, and Naradmuni himself had been recruited earlier into Koirala's Nepali National Congress, he and his followers effectively acted on their own initiative and this can, I believe, be seen as a local rebellion or revolt, exemplifying into the 20th century a continuation of the strong tradition in the eastern hills of resistance to state authority emanating from Kathmandu that had been visible from the very early days of 'unification' in the late 18th century. This rebellion, however, was led by someone with important links with 'the centre' and not just by some peasant or ex-serviceman.

Naradmuni was a well-known and respected figure in Bhojpur as he had established the first English school there in 1947, making good use, ironically, of his connections with the Ranas to secure funds from the government, and had also been nominated as one of two eastern Nepal representatives on a committee supervising the use of a fund for war orphans set up with money from the British government. When Congress was about to launch its invasion from India, Mohan Shumshere Rana wrote to Naradmuni and his uncle Harka Raj, asking for their help in defending the regime, and even after Naradmuni and his supporters had effectively staged a bloodless coup in Bhojpur, the Bada Hakim, not realizing this, handed his own gun over to Naradmuni and told him to fight the rebels!

As the campaign developed, the rebels were able – as the Maoists would do decades later – to capture weapons from government forces and also to rely on substantial local popular support. Whelpton remarks that 'there is evidence, however, of a degree of coercion, as with the later Maoist rebellion: 'proceedings' (*karyabahi*) were threatened against those unwilling to provide provisions for the *Mukti Sena* and, although it was finally decided not to make unarmed civilians accompany the assault force, in the run-up to the second assault on Okhaldhunga in mid-January 1951, the local population was threatened with death if they did not assemble at a nearby crossing on the Dudh Khosi (Karki, 2008, p.102, cited in Whelpton, 2013, p.42).

Talks between the rebels and the Rana government were held in mid-January 1951 in Delhi and a ceasefire was agreed by the Indian government, King Tribhuvan and the Mohan Shumshere Rana regime. By the time fighting stopped in February 1951, Naradmuni Thulung and his fighters had secured a territory that extended from Taplejung on the Sikkim border west as far as Dhulikhel on the eastern rim of the Kathmandu Valley. This was noted with concern in Kathmandu where, apparently, people were worried that 'the *Kiratas* are coming!' (Karki, 2008, p.189). As in all cases of rural unrest in the eastern hills, 'the ethnic dimension' was important, the local peasantry was heavily involved, both voluntarily and involuntarily, but the question as to how far this can be regarded as a 'peasant movement' is debatable. It would perhaps be best to regard it as a rebellion, and a local contribution to the struggle against the Rana regime.

It was, however, a typically local insurgency, with a tendency for individual units to go off in their own direction, even if they were under the overall direction of the high command under Naradmuni. The initial capture of Okhaldunga, for example, was achieved by Mughadan Rai, who gathered armed supporters and effectively acted on his own after failing to maintain adequate contact with the principal leaders of the movement. There were several cases of friction between different groups of insurgents trying to operate against the same target and even an internal revolt within the movement itself, led by Dhan Bahadur Rai, who won over some of the fighters still in Bhojpur to his side, and put both Naradmuni and his main collaborator, Ram Prasad Rai, under arrest on 6 March 1951.

Naradmuni's authority was only fully restored when B.P. Koirala, then Home Minister in the new coalition government, ordered the intervention of J.B. Yakthumba, a leader of the *Mukti Sena* that had been originally organised in India to provide support. The interim Bhojpur government, like the administration set up by Narendranath Bastola and his colleagues in Ilam, continued to function more or less independently of the government in Kathmandu until B.P. Koirala himself negotiated their dissolution during visits to the eastern region in the summer of 1951.

Krishna Prasad Bhattarai, who took part in the armed movement himself, once told an interviewer that the difficulty of keeping their fighters under control was the principal reason why the Nepali Congress accepted a ceasefire in January 1951. Whelpton suggests that 'the crucial factor was probably the pressure on Congress from India to accept the compromise that India had itself proposed, and which King Tribhuvan and the Ranas had already agreed to'; but he also accepts that 'Bhattarai's words underscore the fact that discipline was a problem both within the eastern hills insurrection and elsewhere' (Whelpton, 2013, p.43).

This tendency for local groups to act more or less independently and even to shift allegiance to other emerging dissident groups was a feature of several other movements that developed in the year or so after the overthrow of the Ranas, while the awkward new political dispensation was in play. For example, Mukti Sena members from Bhojpur who were recruited into the Raksha Dal, an auxiliary security force set up immediately after the overthrow of the Ranas, were also at the heart of the brief Raksha Dal revolt in January 1952.

The Far Western Tarai

The armed struggle against the Ranas led to the establishment of a 'revolutionary government' in several areas, including the far west. Thapa states that 'in November 1950, the revolutionary government of western Nepal declared the end of the *bataiya* (sharecropping) and *hunda* systems, and distributed all such lands to the cultivators'.It must be seen as a 'truly revolutionary development', but 'unfortunately, the democratic government formed after the tri-partite agreement signed in New Delhi ordered to return all such lands to the landlords' (Thapa, 1996, p.75).

The 'revolutionary government' was presumably that of Bhim Dutta Panta. Panta had been involved in the Indian freedom movement – having links with Indian peasant uprisings such as Tebhaga, Nijaibole and

Telangana - and, according to Pokharel and Basyal (1999, p.155) was detained in India for 18 months because of this. He had joined the Nepali Congress Party in 1949 and organized and mobilized the local peasants and agricultural workers and became a target for the local landowners and for the Rana government. He was very unpopular with local landlords, who tried unsuccessfully to kill him. He was caught by them on several occasions, but freed because of public pressure.

He was then involved in the movement against the Ranas in 1950, operating firstly in Dadeldhura and Baitadi in the far-western hills and then in Kailali and Kanchanpur in the far- western *Tarai*. In February 1951, he was made governor of the Mahakali region by the new coalition government. He continued, however, to be a committed radical, opposed to all forms of bonded labour, including *haliya* and *kamaiya*, and supported the idea of 'the rights of the tillers'. He raised awareness among the local peasants and workers with the slogan: 'either you cultivate the land or you leave' (INSEC, 1995, cited in Basnet, 2008, p.23). He apparently continued to lead a simple life, even though governor, often taking salt, rice and clothing from the black market and distributing them to the poor. He recruited all kinds of people, including *Tharus* and *Magars* from Kailali and Kanchanpur, to his cause. One of his popular slogans was, apparently, *kita jota halo; kite chhoda thalo; nafra bhaye aaba chhaina bhalo* (either you plough the land or you leave the land; otherwise you will be in trouble).

He left the Nepali Congress, because he could not agree with the party's compromises, as he saw them, and joined the Communist Party after he met Krishnaswami Adkar from India (CSRC, 2008, cited in Basnet, 2008, p.23). Even the new government came to regard him as a 'dangerous radical communist', issuing a warrant for his arrest and a reward of 5,000 *rupees* for his life (ANPA, 2004, cited in Basnet, 2008, p.22). The government of India was also concerned about his activities and his connections. Attempts were made to arrest him in March 1951. He was, however, recognized and supported by Nepali communist party leaders, Pushpa Lal and Man Mohan Adhikari, and he recruited some 500 young militants to join his struggle against the landowning class and for a revolution (ANPA, 2004, cited in Basnet, 2008, p.23).

He attacked the *Bada Hakim* of the region, Mohan Bahadur Singh, for having sponsored the production of fake *tamasuk* (contract) documents, and was arrested and sent to Kanchanpur jail. But he escaped and ran away into the Jogbudha jungle. He was arrested again in February 1953 and imprisoned in Belauri jail in Kanchanpur. His peasant supporters managed have him freed in July, only for him to be arrested again a few days later. Then, it was announced on the radio that he had fled by 'devastating the jail' (INSEC 1995, cited in Basnet, 2008, p.24); and Nepal's army chief, Kiran Shumshere Rana, announced a reward (of variously 500 rupees and 5,000 rupees, according to different accounts-, 2004 and INSEC, 2005) for capturing Bhim Dutta Panta, alive or dead.

It is said that the entire local police force, aided by troops from both the Nepal and Indian armies searched for him. But Panta apparently disguised himself by shaving off his beard, hair and eyebrows and continued to harass the landlords, aided by the locals who protected him. On one occasion, it is said that secret agents discovered his whereabouts and the Indian Army encircled the place on three sides; there was a clash between the Army and Bhim Dutta Panta's supporters, and many were killed, but Panta managed to escape. Spies were deployed, however, and eventually on 1 July 1954 he was found and arrested at Budar village in Doti[23] where he was seeking refuge – allegedly, he was betrayed by the locals who were tempted by the reward offered for him. He was taken to the nearby jungle and severely tortured before being beheaded with a *khukuri* by a policeman.

It is said that he knew he was finished and cried out: 'I wanted to be free from feudalist oppression' and repeated this until his head was cut off. News of his death spread rapidly, and the local landowners celebrated as if for a festival. His head was brought to Khalanga and hung on a bamboo with a note saying: 'your head will also hang like this if you take part in mutiny or treason'; and paraded around. Some 300 of his followers were also arrested and some of these, about 50, were not released until

---

[23] Or, alternatively, in Dadeldhura (Karki 2002 , p.10).

1956 (INSEC, 1995, cited in Basnet, 2008,p.22). Panta's wife and family and friends asked for his head but Mohan Bahadur Singh abused them and forced them to sign a 31 point document committing them to keep the peace. Only then was his head released, and the funeral rites could be held at the temple of Gatal Baba in Dadeldhura (INSEC, 1995, cited in Basnet, 2008, p.24).

The West Central Tarai

In Lumbini, a peasant movement directed against the landlords and called simply *jamindar birodhi andolan* that developed in the early 1950s, was closely linked to the activities in the area of the Nepali Congress. Before November 1950, Singh, the son of a minor Thakuri landowner, who had been educated in India and had served for some time in the Indian army, where he had acquired some medical training, set himself up as a doctor based at Nautanwa, just south of the Nepal-Indian border, He travelled extensively in Nepal's western *Tarai*, ostensibly to treat patients. After meeting Bholanath Sharma and other Congress activists, he became committed to the cause and proved an effective political organizer, making use of his contacts through his medical work and links to other ex-servicemen. He still operated out of Nautanwa, where he became manager of the Congress Party office.

He had previously clashed with local *zamindars,* who were particularly unpopular with the peasants and who had, allegedly, brought in thugs (*goondas*) from India to attack and loot the villages. Singh promised that Congress would now ensure they were punished. Once the Congress armed movement began, the *zamindars* moved their families and moveable property to India and also cooperated with the local *Bada Hakim* against K I Singh and his forces. The struggle was vicious, as the *zamindars* were ill-regarded, and several were killed (Whelpton, 2013, p.45).

When hostilities against the Ranas started, Singh led a small force of a dozen or so men, with eight or nine rifles, across the border and set up camp at the village of Myudihawa, switching later under pressure from government forces to Karabla, which was more defensible. He managed to obtain additional weapons and food from the local peasantry, and was joined by a combination of student volunteers and ex-Indian army colleagues. Singh refused to accept 'the Delhi compromise' and ignored Congress' call for a ceasefire. He launched a further attack on Bhairahawa and maintained his rebel base and fighting force. His force had reached about several hundred when he was eventually arrested and jailed in Bhairahawa in February 1951, following a combined operation by forces deployed by the new government combined with Indian troops (roughly 1000 against 200-400).

Singh escaped from jail, was re-arrested and sent to Kathmandu. He was released from prison together with two leaders of a *Kiranti* secessionist movement in January 1952 by mutinying members of the Raksha Dal, a militia composed of former members of the *Jana Mukti Sena*. He fled to Tibet with some of his supporters, declining the offer to make him leader of the revolt, and the authorities were able to restore control without external intervention by Indian troops. The government then disbanded the *Kiranti* sections of the Raksha Dal and, on India's suggestion, banned the Communist Party of Nepal, which had been sympathetic to the revolt.

In May 1952, however, agitation developed in the Nautanwa-Bhairahawa area, led by K I Singh supporters, who claimed to be influenced by the Chinese communist experience, even if he was still linked to the Nepali Congress. Singh later returned to Nepal to form the United Democratic Party, possibly with covert funding from the king to help counter what he felt was the dangerous dominance of the Nepali Congress. He served briefly as prime minister, appointed by the king, in 1957. Whelpton dubs him a 'royalist Robin Hood', and it seems likely that he did receive support from the king; it seems that Singh saw the possibility of an alliance of the monarchy and the radical forces against the social democratic or centrist Nepali Congress - much as did the Maoists at certain times during their People's War, as argued by M B Singh (2003)[24].

---

[24] How far Mahendra himself countenanced this possibility is a moot point, but it is perhaps significant that the king decided in 1956 to re-legalise the Communist Party (banned in 1952 for complicity in the Raksha Dal revolt), and

The Western Hills

Communist influence was also beginning to grow in the western hills during the early 1950s. A leading activist in this regard was Mohan Bikram Singh, later to become a key figure in the Maoist movement in Nepal[25]. Singh was born in Kathmandu in 1935, but his father was a wealthy landowner with an estate in Pyuthan District in the western hills[26], and it was there that Singh grew up. He moved from Pyuthan to Kathmandu after finishing school and started reading left-wing literature; he was particularly influenced by political leaders who had started to emerge in India, notably the communist Rahul Sankrityayan; but joined the Nepali Congress, and took part in their actions against the Rana regime. His father also became a leader of the anti-Rana movement in Pyuthan during the early 1950s, but his father's motives as a rebel were anything but revolutionary, and his reputation as a renowned 'feudal' was considerable, even as far away as Thabang in Rolpa. In fact, he had notoriously increased his landholding by making local peasants sign false contracts.

M B Singh then joined the *Rastriya Praja Parishad* (National Democratic Party), shortly before moving to the Communist Party in December 1953. He and Khagu Lal Gurung organized a three-month training programme or 'progressive study group' in Ratamata near Dankhakwadi village in their home district of Pyuthan, with the objective of instilling the precepts and principles of communism in any of the local peasants and workers who signed up; the programme also included physical education. News of this spread, and soon some 150 participants were involved. In February 1954, as the programme drew to an end, a delegation of residents from Narikot to the north of Dankhakwadi, came to see the two leaders, complaining of the abuse they had received from

---

apparently also funded the Marxist-Leninist faction (the Ma Le) which was the continuation of the ultra-radical 'Jhapeli' group.

[25] This section draws heavily on the work of Benoit Cailmail (2008-09, 2013).

[26] M B Singh referred to his father as 'a feudal reformist' as he was against corruption and for democracy but against all kinds of socialist and communist ideology (Cailmail, 2008-09,p. 17).

their local headman (*mukhiya*) and his allies, who had mis-appropriated local peasants' land by forcing them to sell it.

Accompanied by hundreds of militants recently converted to revolutionary principles and armed with sickles and cudgels, Singh and Gurung went to Narikot to persuade the *mukhiya* to return the land. After nearly a week of fruitless talks and even after some 600 peasants had mobilized, holding meetings and organizing demonstrations and processions, the landowners remained adamant, arguing the legality of the transfer of property and showing the documents signed by the persons concerned – a trick Singh's own father had played on peasants himself. Singh, therefore, opted for more radical methods: he kidnapped the main landowners and took them to Machchhi (Okharkot), his stronghold, where they would stand trial in a people's court. Realising the need to counter this decisive move, the remaining landowners of Narikot, accompanied by landowners from other villages, notably Khung, attacked the revolutionaries at Machchhi by night, with sticks, knives (*kukhuris*), swords and even rifles.

Singh and Gurung were prepared for this, however, and ambushed the attackers at the entrance to Machchhi. Armed themselves with sticks and stones, they also set their dogs on the attackers and forced them to retreat. The landowners were obliged to return the documents justifying their forced purchases of land and on 7 March 1954, the documents were destroyed. The immediate aim of the action taken by Singh, Gurung and others was a form of 'restorative justice', but in other actions taken in the same period, Singh advocated the transfer of 'land to the tiller' – a typical Naxalite slogan.

With this first victory of the communist party in Pyuthan, the 'revolutionaries' set out for the nearby villages of Bangemarot, Badikot and Tusara to spread the word about what was being claimed as a precedent. Singh and Gurung launched new protests and demonstrations during 1954, supporting peasants' rights and fighting corruption, as well as calling for the transfer of the district headquarters from Khalanga to Bijuwar. Singh even called on his own father – unsuccessfully - to restore the land he had taken from 'the tillers' to their rightful owners, an act that endeared him even more to his followers. After his father died, and Singh inherited the land, he returned the land that had been taken illegally from the peasants to them.

The peasant movement was remarkably effective within Pyuthan and particularly within the areas of Okharkot, Narikot, Bangemarot and Tusara, but appears not to have spread much beyond. As regards the leadership itself, Benoit Cailmail quotes Bianco on the Chinese experience: 'the leaders [of the peasant revolts] are more frequently non-peasants than peasants, and among the latter, mostly well or fairly well-off farmers, than poor peasants' (2005,p.149), and suggests that this is also the case in Nepal, at least in Pyuthan in the 1950s. He concludes that the leaders of the peasant movements in Pyuthan tended to share relatively comfortable backgrounds – although he recognizes that Khagu Lal Gurung was a peasant, farming just over 1 hectare of land throughout his political career (Cailmail, 2013,pp.144-147) – and had all attended school at a time when that was still exceptional.

He accepts that such generalization is difficult, and in any case, this was not the case for most of the grass-roots activists who made up the cadres and committed supporters of the movements. We shall see that similar movements, albeit on a more localised scale, were encouraged in the 1970s by young outsiders with political commitments and attachments who established a library and training for local cadres, and led a peasant movement against local landowners and government functionaries – the Chintang Bidorha (1971-1979). They too were to experience a heavy response from local elites, leading to the arrest and detention of the identified leaders (see below).

Eventually, in January 1955, while leading demonstrations in Bagdula, Singh and Gurung were surrounded by police and arrested; they were sentenced to two years in jail. Even from jail, Singh and Gurung continued to agitate for people's rights and revolutionary change. They were transferred from prison to prison, the last transfer being after they launched a rebellion inside the prison in Palpa where they were incarcerated to object to prison conditions. The wardens tried to suppress the rebellion by beating up Singh, but he fought back, striking the guards with his shoes. When the other inmates rallied round him, it was decided that, to avoid a prison riot, he and Gurung should be transferred to Salyan. When news of this event reached Pyuthan there were large demonstrations demanding his release.

He was released in 1956, and the next year was elected to the Central Committee of the CPN. Two years later, in the first democratic elections held in Nepal in 1959, the CPN secured exceptionally strong support in the area where Singh had earlier mobilised the peasantry. Indeed, the legacy of Mohan Bikram Singh and Khagu Lal Gurung in Pyuthan in particular was immense, as Benoit Cailmail notes, and remained strong throughout the *Panchayat* Era, through the early 1990s and during the Maoist insurgency (2013, pp.144-161).

Bakhtapur, Kathmandu Valley

In Bhaktapur, peasant unrest broke out in Somlingtar in 1951 with a focus on the rights of tenants and the non-payment of grain (*kutbali*)[27]. There were other, related instances of unrest in Kathmandu and Bhaktapur later the same year. The demands of the peasants included the abolition of landlordship and the *talukdari* system. These actions by local peasants led to the formation by the Communist Party of a peasant or farmers' association – the *Akhil Nepal Kishan Sangh* (the All Nepal Peasants' Association - ANPA) – whose main concern was to combat the control exterted over the peasantry by *birta* owners and *zamindars,* and which advocated the 'rights of the tillers'. The ANPA – popularly known as *Kisan Sangh* - rapidly developed to become involved in peasant struggles in various parts of the country, in the *Tarai* as well as in the hills, according to Padma Bahadur Budhathoki, one of the founders of the association (Budhathoki, 1981, p.3).

According to Basnet et al. (2010,p.119), the refusal of tenants in Somlingtar in Bhaktapur to pay rents in the form of grain payments 'set off a movement against share-cropping in Rajapur, Bardiya District; the

---

[27] Under the *kut* system, the cultivator paid a stipulated quantity of food grain or other commodities, or a stipulated sum in cash.

jamindar birodhi andolan in Lumbini; the *dharma bhakari andolan* in Bara and Rautahat Districts; led to an organized and focused movement against the *birta* tenure system and the *jamindari* system, and in support of tenancy rights; (and) inspired the formation of an agricultural union parallel to the *Akhil Nepal Kishan Sangh'*. This refers, presumably, to the *Krishi Sudhar Sangh*, formed by Basu Pasa, a peasant leader from Bhaktapur, who accused the ANPA of being controlled by the communists. But the two associations had similar objectives and concerns, and Budhathoki states that the All Peasant Association Reform Committee was formed on 8 February 1955, with Budhathoki as president and Basu Pasa as secretary (Budhathoki, 1981, p.8).

According to Karki (2002, p.10), 'this helped to mobilise peasants of Kathmandu Valley who engaged in the struggle, demanding that the *guthi* should accept payment of half cash and half kind instead of claiming two thirds in rice and one third in cash. They further demanded that the date of paying the crop contribution be extended up to the middle of March. The action policy and objective of these struggles was explained in a 12-point programme, including the distribution of public land to local landless people. According to Basnet, in October 1954, there were demonstrations by peasants in Kathmandu and Bhaktapur against the government, which produced six demands, including the abolition of the landlord system. Demonstrations were held and some 19 protestors were arrested. This paved the way for a second phase of the revolt in March 1957, following which 55 peasants were arrested and detained (Basnet et al. 2010, p.119).

These peasant movements in Bhaktapur continued, according to Thapa (2001), in one form or another, until the 1980s; and, certainly, the formation in 1975 of the Workers and Peasants Party under the leadership of Narayan Man ('Comrade Rohit') Bijukche both reflected and contributed to the continuing importance of the peasant movement in Bhaktapur up to the present day. Bhaktapur remains a 'special case', and a history of the specifics of the peasant movement there, and of the Workers and Peasants Party, is long over-due.

The Eastern Tarai

The *Tarai* Congress was founded in 1951 (as the *Tarai Newasi Sangh* – Association of *Tarai* People) by Kulanand Jha, Ram Janam Tiwari and Syam Lal Mishra after they failed to get assurances regarding provincial representation in any future Nepal Congress government. In 1953, the party claimed 60,000 members and demanded proper representation of the *Tarai* in the proposed Advisory Assembly. It sought recognition of Hindi as an official language, the establishment of an autonomous *Tarai* state, and adequate employment of people from the *Tarai* in the Nepali civil service. In 1954, the leadership passed to Kulanand's brother, Vedananda Jha, who was appointed as Minister of Home Affairs, Development, Industry and Commerce in 1961 and later (in 1977) became Ambassador to India.

In Bara and Rautahat, the peasant movements were focused on *dharma bhakari* – best described as 'a food grain fund', which involves local contributions and is made available to those in need either at the beginning of the agricultural season or during the 'hungry period'. It is not entirely clear what the issue was, but the movement in both districts was referred to as *dharma bhakari andolan* (the *dharma bhakari* movement).

In early 1952, there was a revolt by poor and landless peasants in Rautahat and a movement to destroy loan contract papers drawn up to legitimize high interest loans made by money-lenders and large landowners to these people. Known as the *Tamsuk Fatta Andolan* (the bondage paper destruction movement), its attacks were directed not only at those individuals considered to be guilty personally but also at district land registration offices where the records and land mortgage papers (*bhog bandagi*) that were often used as security against the loans were held (Kishan Sangharsha, 1977, p.30). The movement was crushed, with numerous arrests and detentions made by the police.

Despite – or perhaps even because – of state repression, this was followed in 1952-53, by a movement in Rautahat and Bara which struggled not just for land and other rights, but for dignity and social respect. It was called:'re nahi, ji kaho andolan – The Say 'Hello', not 'Hey you' Movement – and protested against the use of disrespectful and humiliating language by landlords and by bullying of peasants and landless labourers. According to Karki (2002 ,p.10), 'it played an important role in the history of peasant movements in Nepal' and had an influence on subsequent movements. In Birgunj, where there was a procession with slogans, five were killed and 125 injured; in Rautahat, Asarfi Saha was murdered in 1953. The peasants involved also adopted the slogan '*maddat bhakari, khada karo*', and helped each other to save and store grain in support of their struggle (INSEC, 1995, p.158, cited by Basnet, 2008, p.25).

At the beginning of 1953, there was a rally of the All Nepal Peasants Association (ANPA) in Bara, calling for the land of the big landowners and big *birta* holders to be confiscated and distributed to middle and small farmers and landless people. It called also for an end to exploitation of small farmers and landless people by feudal landlords. The landowners rallied to suppress the movement. According to the ANPA chairman, Keshav Badal, 1,000 kg of rice was cooked at each mealtime for the participating peasants. This huge gathering of protestors was attacked by armed police under the regional administrator on 22 January 1953 to arrest the peasant leaders: Madhu Singh, Comrade Ran Vriksya Chaudhari, Nanda Kishwar Chaudhari, Mohan Singh Mahato, Nandan Aryal, Dev Narayan Kalwar, Ram Lal Tharu, Rama Kanta and Nathuni (Karki, 2002, p.11). During the struggle, 19 year old Mukha Lal Mahato was killed by police fire. Others were arrested and tortured; and the movement was suppressed.

On 15 May 1953, the Nepali Congress Party launched a 'no-rent' campaign in the eastern *Tarai* at Biratnagar where B.P Koirala addressed a crowd of 35,000 people asking them to defy the government's 'lock-out' order regarding Biratnagar Jute Mill. During the first week of June, the movement spread to the Birgunj, Biratnagar, Saptari, Janakpur, Mahottari area. Ram Narayan Mishra along with Saroj Koirala toured the region organizing peasant meetings and asking them not to pay their land rent.

Between 1953 and 1956, there was a movement which started in Jhapa and moved to other regions of the *Tarai*, called *khamar roko* or *khamar rok* (Abolish Owner's Private Property), which seems to have been influenced by communists in Bengal. Khamar in Bengali means *nij-jote* land, which is equivalent to *jirayat* land. It seems that this movement was led by young farmers (Thapa, 2001, cited in Basnet, 2008, p.26). The government formed 'a small reformist commission' in response to this movement, but this did not represent the tenants or landless peasants in any way (Basnet, 2008, p.26). In 1955-56, a movement in Siraha resulted in the death of Bahadur Saha. Between 1956 and 1958, there was a peasant movement in Saptari, in which Aghori Yadav was killed (in 1958).

The Western Tarai

A land rights movement – the Beluwa Banjari Peasants Movement – began in Dang Deokhuri District in December 1959, inspired by the Nepali Congress Party's 1959 election slogan - *'jagga kasko? jotneko. ghar kasko? potneko* ('To whom does the land belong? To the tillers. To whom does the home belong? To the tenants'). This movement was directed against landlord 'incursions' and evictions, and involved mainly *Tharus*. It developed further during 1960. In an effort to contain it, one of the leaders of the movement, Gumraha Chaudhury, was murdered. The death of Gumraha further aggravated the conflict (Thapa, 2001, cited in Basnet, 2008, p. 26).

Tharu women were prominent in this movement, notably Lohani Chadhury and Somati Chaudhury; Ratna, Kalu and Lahani were arrested and tortured, but released after a week. 'When the landlords came to loot mustard in 1959, all of us - except old people and children - went out to defend ourselves with sticks in our hands. Lahani suffered injuries to her back. But we were not hopeless; we did not leave the ground to the landlords. Rather, we chased them and their servants away. When they came again next morning, we chased them again', said Somati Chaudhury. 'They came back again in 1960. The struggle continued throughout the month of June-July. Police assisted the landlords; they haunted the area constantly to look for people. One day, Ratna and Kalu were arrested. Gumraha was murdered. We had to fight hard to save the villagers from the *zamindars* and the police'.

Peasant Movements in the Panchayat Era

When King Mahendra dissolved the multi-party system that had been in place during the 1950s, in favour of a party-less *Panchayat* 'Democracy', all political parties were banned and their leaders threatened with imprisonment. The communists officially moved to India, first to Darbhanga in Bihar, then in Varanasi from 1961 to 1974, and finally to Gorakhpur; but the activists remained inside the country, operating underground, holding meetings, distributing pamphlets and even leading or participating in a few demonstrations.

The All Nepal Peasants Association (ANPA) was banned and a state-sponsored 'Nepal Peasants' Organisation' was established to 'represent' the interests of small farmers. Mahendra initiated various programmes of rural development; and one of the most significant attempts at far-reaching land reform was made with the Lands Act in 1964 – a government initiative whose significance has been much debated over the years (see Adhikari, 2008), but which undoubtedly gave an impetus to increased rural unrest.

The Western Tarai

A peasant uprising took pace in Beluwa-Banjadi in Dang Deukhuri in 1961. As a result, the landlords in the area began to deprive substantial numbers of tenants, who had been cultivating the land on condition of assured tenancy rights, of their entitlement. Instead, they attempted to impose sharecropping arrangements giving themselves rights to 50 percent or even in some cases two-thirds of the harvest. In response to this, peasants from Gairagaon, Gulariya, Ratanpur, Beluwa-Banjadi and Karjahi organized themselves to resist the landowners' initiative. They were aided by Communist Party activists Narayan Prasad Sharma, Ram Prasad devkota, Manik Lal Gautam and Keshav Raj Gautam. On 31 January 1961, the landlords recruited hired thugs to attempt to seize the mustard crop from the threshing grounds of the peasants of Beluwa-Banjadi. A serious clash resulted, during which the peasants managed to capture some of the hired men and handed them over to the local administration. According to the local ANPA leaders, the landlords made no further attempts to seize the peasants' crops (Karki, 2010, p.11).

In the years immediately following the Land (Reform) Act of 1964, there was a rush of title transfers in many parts of the *Tarai* as landowners tried to devolve themselves of official ownership. In many cases, tenants and sharecroppers were replaced by cultivators in whose name the land was registered, but who remained effectively subordinate to the original landowner; in other cases, *ukhada* tenants were replaced by daily wage labourers. This latter tendency was particularly prevalent in the western *Tarai*.

In September 1966, a crowd, estimated around 4,000 surrounded the district administrative office in Parasi in response to fears (and the reality) of such changes. Local security forces opened fire, killing some and injuring more (the official figures were nine and 12 respectively). No official investigation was ever carried out. Although this was clearly a local peasant demonstration, it was labelled by the Kathmandu media, notably the state controlled *Gorkhapatra,* as having been instigated by 'certain elements' who had 'misguided the peasants on the issue of land reform' and who should 'not be left unpunished'. These 'elements' were allegedly linked to India.

There was another shooting incident in Nawalparasi after the car of the assistant Minister of Education was stoned and the chief district officer attacked, as police fired on the crowd and four people were injured. Some 'egregiously offensive local officials were removed and arrested students were released as in Kathmandu', but 'firm suppression of people was applied at a relatively early stage, with the army sent down to control the border and to suppress the agitation. Police were given 'shoot at sight' orders to suppress the agitations' (Sah, 2017, p.286).

According to the blueprint of the compulsory savings scheme initiated by the government in 1965, every farmer was required to deposit a small percentage of his crop with the government each year. This was viewed by most farmers as an additional tax, despite the government's promise to return the deposit with interest at the end of a five year period. Many of the larger landowners were able to avoid making these compulsory savings; but small farmers and peasants were less able to resist or refuse, and savings were often collected using various means of coercion or intimidation from those least able to afford to give up any portion of their income. Over the next few years, there were also constant reports in the press suggesting that the officials concerned were not keeping adequate records of the amounts 'saved' in this way and that the farmers' savings were being, in effect, misappropriated.

Resistance to payment gradually increased, and in April 1969 broke out in Taulihawa, the capital of Kapilvastu District when a crowd of women surrounded and attacked officials attempting to collect compulsory savings. Violence quickly spread across Kapilvastu District and into the adjacent Rupandehi District. Compulsory-savings deposits were broken into a looted and burned. The police responded, often violently with shooting at protestors and looters in some thirty villages over a three-week period, as a result of which some 33 people died, hundreds were injured and many others arrested. The movement forced the government to suspend the scheme for compulsory savings throughout the country.

Divisions within the Communist Movement

After King Mahendra's coup, when all political parties were banned, the Communist Party moved to India – first, to Darbhanga in Bihar, then to Vanarasi from 1961 to 1974, and finally to Gorakhpur--but many activists remained inside Nepal and continued the struggle underground. In 1961, when the party divided into different tendencies or factions, K I Singh rallied the leftist groups, demanding a Constituent Assembly. This received considerable support amongst the party grassroots, but he did not get much support from fellow Central Committee members. When the party eventually split in 1962, Singh sided with the more radical Communist Party led by Tulsi Lal Amatya and maintained a revolutionary commitment to popular mobilisation and armed struggle 'when the time was ripe' throughout his subsequent political career.

In 1962, at the Third Party Congress, the Communist Party of Nepal divided effectively into a pro-Soviet and a pro-China faction. In 1968, Pushpa Lal, the founder of the Communist Party of Nepal (CPN), also left the main body of the Communist Party of Nepal (Tulsi Lal Group) to establish a separate group, which became known as the CPN (Pushpa Lal). More communist groupings developed from this original split. In 1971, a group of CPN leaders — Manmohan Adhikari, Sambhu Ram Shrestha, and Mohan Bikram Singh — were released from jail and together formed the Central Nucleus, which tried to unite with Pushpa Lal's group. This failed, because of Pushpa Lal's objective of uniting all the democratic forces, including the Nepali Congress, against the *Panchayat* Regime, and Central Nucleus itself broke apart[28].

The Eastern Tarai

In 1970, what started as a local struggle by settlers (*sukumbasi*) in the area of Biratnagar to occupy an area of forest land (known, ironically, as *ramailo jhoda* or lovely Jhoda), intensified and began to spread. In January 1971, the army encircled and opened fire on landless settlers who were clearing land in the forest in various parts of Morang District – in Kanepokhari, Pathari, Dahijhoda, Jarayotar, Bayarban, Jadaha, Bhaunne and Chisang. The *sukumbasis* were defenceless against the bullets of the army; those killed were allegedly loaded onto trucks and later thrown into the Koshi River. There was a national protest by *sukumbasis,* supported by the banned political parties in outrage against the killing. As a result of this, the government introduced the *Jhoda* Land Act 2028 (1971) which guaranteed land rights to landless people up to a maximum of four *bighas* per family. This Act has, however, never been fully implemented.

---

[28] Adhikari formed the Communist Party of Nepal (Manmohan), which developed close relations with the Communist Party of India (Marxist); M. B. Singh and Nirma Lal founded what became known as Communist Party of Nepal (4th Congress) after the Fourth Party Congress held in September 1974; other splinter groups included the Nepal Workers and Peasants Party, the Communist Party of Nepal (Krishna Das), the Communist Party of Nepal (Burma), and the Communist Party of Nepal (Manandhar). In 1975, Pushpa Lal's group developed into the Communist Party of Nepal (Marxist-Leninist) and the original CPN became just one of many factions, eventually taking the name CPN (Amatya) and adopting a pro-Soviet position.

According to Karki, another military operation in 1976 displaced the *sukumbasis* from Jhoda, situated to the south of Chulachuli and to the north of the East-West Highway. Again, in 1983, the entire village of Chulachuli was displaced, following a declaration that the land was to be used for re-forestation. On 8 January 1985, as many as 2,400 men and women marched towards Bahragoth, Chulachuli, carrying photos of the King and Queen; they were *lathi*-charged by some 400-500 police near the Ratuwa Bridge on the East-West Highway. Many protestors, including women, sustained serious injuries. The issue of Chulachuli remained unresolved throughout the *Panchayat* period (Singh, 1992, pp.338-353, cited in Karki, 2002, p.12).

In 1971, under the influence of the teachings of Mao Tse Tung and the experience of Charu Mazumdar, architect of the Naxalite uprising in Bihar and elsewhere in India, a group of young communist party activists – including K.P. Oli, Radha Krishna Mainali and Chandra Prakah Mainali - formed the Koshi Regional Committee of the CPN, in Jhapa. These young activists launched an underground guerrilla movement, popularly known as 'the Jhapa uprising', in line with the Maoist concept of a protracted People's War. Pandey and Rimal draw attention to the fact that several women were closely involved in the 'Jhapa Bidroha', including Sharada Mainali, Lila Kattel, Gaura Prasain and Sita Khadka, providing shelter to the underground leaders and 'exchanging news and information'; three of these women served jail terms as a result (Pandey & Rimal nd, p. 17).

The movement actually started in Jymirgadi village in Jhapa on 16 May 1971 and attracted the attention and support of many young activists across the country. It was directed mainly against the large 'semi-feudal' landowners and led to the killing of eight landlords. The movement was brought to a swift end by a brutal counter-insurgency campaign by the police, which led to hundreds of its followers being killed, jailed or forced into exile. Five leaders of the movement - Ramnath Dahal, Netra Ghimire, Biren Rajbanshi, Krishna Kuinkel and Narayan Shrestha - who were subsequently executed (on – variously – 24 February or 4 March 1973) by firing squad in a wooded area in Sukhani, Ilam after being taken there from Chandagadhi, Jhapa under the pretext of shifting them to another prison (Karki, 2002, p. 12). K. P. Oli was also arrested, was released shortly but was then again arrested and charged with treason. He spent 14 consecutive years in jail until 1987[29].

Despite the crushing of the movement, members of the group continued to conduct clandestine political work amongst the peasants of Jhapa and, in 1975, the survivors of the Jhapa movement took the initiative to found the All Nepal Communist Revolutionary Coordination Committee (Marxist-Leninist). Other small groups merged with ANCRCC (ML) and, on 26 December 1978, ANCRCC (ML) organized the founding congress of the Communist Party of Nepal (Marxist-Leninist). C P Mainali was elected general secretary of the party. CPN (ML) – known as Ma-Le - was an underground party, and during 1979 conducted small-scale, armed activities against the regime and feudal landlords, with a view to initiating 'an agrarian uprising' (according to Deepak Thapa (2001), cited in Karki & Seddon, 2003, p.12).

The tactics of armed struggle, however, did not prove particularly successful, and the party changed its political approach, and started to focus more on mobilizing mass movements for a democratic change. Mainali, clearly identified with the initial militant phase of the party, was removed from his post of general secretary and replaced by Jhala Nath Khanal. Subsequently, according to Mikesell, 'abandoned the people's war for underground educating and organizing in village, town, school and campus in order to build a mass movement' (Mikesell, 1999, p.100). The Party united with a number of other communist party factions in the early 1990s to form the Communist Party (United Marxist-Leninist) or UML.

---

[29] In March 2017, now Chairman of the CPN-UML, he led a 15-day campaign which included a visit to Sukhani Park where tributes were offered to 'the martyrs' of the Jhapa uprising, to officially launch the nationwide campaign.

The Western and Central Hills

A local movement began in Barre, in Argha Khanchi District in 1974. Under the leadership of Keshar Mani Pokhare, local peasants launched an initiative to seize property from the local landlords. The government's response to this movement, known as 'the Barre Movement', was rapid and effective. The movement was quickly suppressed (Adhikari, 2008, p.32,Rawal, 1990, p.70).

According to Adhikari (2008, p.32), another outbreak of peasant unrest occurred in 1978 in Piskar village in Sindhupalchowk District when poor peasants, who were mainly *Thami* people, seized rice from the granaries of local landlords. The landlords called in the police to suppress the unrest, and two people were killed. Others suffered as they were forced to flee their village (Adhikari, 2008, p.32). Basnet also refers to what he calls the 'Piskor movement' that he says arose in Sindhupalchowk in 1983. This was directed primarily against the Pandeys of Piskor (Piskar) village. Peasants had been obliged to provide free labour services to these Pandeys, pay high rates of interest on loans contracted with them, and had to hand over 'most of their production' to the landlords (Basnet, 2008, p.27). They organized themselves and demanded *'jasko jot usko pot'* – land to the tillers (ANPA, 2004, cited in Basnet, 2008, p.27). But the movement was crushed and several poor peasants and landless people were killed. It seems as though these two accounts relate to the same incident, but further research is needed to be sure.

The Eastern Tarai

At the beginning of 1978 (or alternatively in 1979-80), there was a peasant uprising in the vicinity of Janakpur in Dhanusha District, when poor peasants and landless labourers launched a movement called *bhakari phor andolan* (Break the Granary Movement), which, among other things, took over the granary of Justice Prakash Man in Dhanusha. But this was relatively short-lived, and was rapidly suppressed by the army (Karki, 2002, p.13, Adhikari, 2008, p.32). Alternatively, the struggle started in Dhanusha but spread to different parts of the country (Thapa, 2001, cited in Basnet, 2008, p.27).

On 6 April 1979, students demonstrated in Kathmandu against the execution of former Prime Minister of Pakistan Zulfikar Ali Bhutto, but as the crowd approached the Pakistani embassy, they were stopped by police at Lainchaur. In the following days, there was student agitation at several colleges in the eastern *Tarai*. But although this started with students, things soon took a different turn and became more widespread. Student protests were soon overshadowed by those by factory workers and by farmers opposed to higher land taxes, and more generally by those responding to corruption and mal-administration. The main centre for this was Janakpur, where the chief district officer and the zonal commissioner were both subsequently suspended.

In May 1979, a variety of protests broke out, still mainly in the area of Janakpur, and there were major clashes which led to the deaths of five people (one policeman and four demonstrators), according to government reports. A Home Ministry report released on 19 May indicated serious disturbances in four separate incidents in Dhanusha District, in Siraha and in places west of Janakpur. The report spoke of 'a series of mob attacks' on customs posts and government warehouses (*godowns*), involving the burning of some buildings, direct attacks on government officials, and attempts to disrupt traffic along the East West Highway.

On 20 May, it was reported the demonstrators had removed portraits of the King and Queen from office and that they had been burned in effigy. The railway terminus at Janakpur was also set on fire, putting Nepal's only rail service out of operation. The independent daily, Motherland, carried a Home Ministry report saying that six people were killed when police fired on a crowd between Janakpur and Birgunj, in Sirlahi District. The next day, although Janakpur was reported quiet, in the neighbouring district, Mahottari, police fired on a crowd resulting in one death and one person injured. Indeed, there were reports of disturbances across the *Tarai* – with what one source describes as 'burning, thrashing, destruction of bridges, barricading of roads' (Sah, 2017, p.285). There can be no doubt that peasants from the region were involved, but this was more than a classic peasant uprising.

The West Central Tarai

In 1977, there was a local movement to seize the property of landlords in Jugedi, in Chitwan District. It was suppressed by the police (Adhikari, 2008, p.32). In May 1979, some 500 *sukumbasis*, who were squatting on public land, assembled in Barghat in Nawalparasi District and surrounded the local police station to protest against the detention of some of their fellow settlers, the seizure of their farm implements and other property. They had previously requested the local *panchas* (village council chairmen), the Chief District Officer and a minister to tour the district and learn for themselves how far the local people were being intimidated and harassed by the forest guards and the local police, and asked that they be accorded their legal title to land. According to Krishna Ghimire (1992, p.144), 'the official response to the *sukumbasis* and their demands was immediate repression... the police opened fire without warning'. Hundreds were injured.

Similarly, in the early 1980s, *sukumbasis* in Nawalparasi continued to resist the indiscriminate eviction of large numbers of households in Nawalpur. A large demonstration was organized and held in Kawasoti *Bazaar* in Nawalparasi on 14 March 1983. Police fired on the demonstration and several people, including the president of the district *panchayat* (who supported the local land rights movement) were arrested and detained in Bhairahawa jail (Ghimire, 1992, pp.145-146).

Karki reports that 'from the early 1980s until 1990, a series of similar demonstrations took place throughout the *Tarai* belt of Nepal, attempting to establish land rights, mainly concentrated ... in Chitwan, Nawalparasi, Banke, Bardiya, Kailali and Kanchanpur Districts' (2002, p.13). We do not have details of all of these local peasant land rights movements in the *Tarai*, but Conway and Shrestha (1981, p.226) considered them politically significant, while Kaplan and Shrestha referred to the '*sukumbasi* movement in Nepal' as 'the fire from below' (1982, p.75), while Ghimire makes it clear that government concerns about the 'problems for law and order' posed by 'large and growing illegal settlement' as dating back to the mid-1970s (Ghimire, 1992,p.5) and traces the evolution of the evolution of 'popular discontent' and of 'official responses' to local demands for land from as far back as the 1960s (Ghimire, 1992, pp.130-146).

The Eastern Hills

In the winter of 1979, in Chintang, in the Khalsa area of Dhankuta District in the eastern hills, there was a local uprising of peasants, mainly *Rais,* against local landlords and authorities, also mainly *Rais.* This was a good example of the situation Whelpton described as one form of political conflict in which 'tensions between the peasantry and those claiming a share of their produce as rent or taxation could result in violence and, where both cultivator and landlord were from the same ethnic group, this could be seen as straightforward class conflict'.

There was a long history of leftist political organization in the area, from the 1950s onwards. In the mid-1960s, Mohan Chandra Adhikari had come to the Khalsa area as a school teacher in Chungbaang, and in 1967, he formed a Khalsa Area Committee as a branch of the CPN-Marxist. The members of the committee were active in promoting concern among the local peasantry regarding the dominance of the *majhyias* in Chintang and surrounding areas of Khalsa. A *majhiya* was in effect a local form of *jimindar*, as the *majhiyas* were appointed as village representatives to collect land taxes and were able to call on local labour to cultivate their own fields on land that was tax-exempt. They also had links to higher level authorities. Rather than being representatives of their community, however, they acted more as state functionaries and as owners rather than custodians of the land although they claimed they were *kipatiya (kipat* holders). In fact, however, land tenure in Khalsa did not share the *kipat* system characteristic of much of the eastern hills, as it was not communally but privately owned.

The Khalsa area comprised mainly hilly rain-fed *bari* land, suitable for millet, maize and legumes; rice production was undertaken only by a small number of farmers in the lowlands. Land distribution was very unequal and many small farmers were also share-croppers, agricultural labourers and porters. Most suffered from food shortages and maintained close serf-like relations with the larger landlords to enable them to borrow food-grains during the 'hungry season'; this meant sometimes that they were obliged to provide free labour (*jhara*) to these landlords (the *majhiyas*) during peak agricultural seasons, and on special occasions.

In 1970-71, two young men, Govind Bikal and Bam Devan, arrived in the village as school teachers[30]. The two young men had links with communists in Dharan, where they came from, and had been influenced by the Jhapa uprising and the Indian Naxalites. They set up a library at the school, giving it the name of Lali Guras Pustakalaya (Rhododendron Library); they collected 'progressive' literature (including the works of Mao Tse Tung) and the library became a centre for local youths, particularly for the discontented. They and other local activists maintained close links with the Party in Dharan, but the library now became a training centre for the Purba Koshi Prantiya Committee, another faction of the communist party. Also, Man Mohan Adhikari, who was now leading yet another communist faction party faction which had split from Mohan Bikram Singh's 'Nucleus', came with other comrades to visit Chintang to train their cadres and establish a party committee.

It sees that there was a serious drought in 1970-71, local food production had collapsed, and many families were starving. The granaries in the *Majhiya* households, however, were full. A 'movement' broke out to 'borrow' grains from these rich households; apparently old women were involved (Adhikari, 2008,p.32). It was not exactly a seizure, and those 'borrowing' the grains even provided receipts; but it was effectively forced on the dominant *Majhiya* households who, not surprisingly, took exception to this action. Some of them filed a complaint against those whom they identified as the instigators, which included Bikal, Devan and other activists, and some of the peasants taking the 'loans', accusing them of theft. The Chief District Officer (CDO) heard the charges and many were arrested and jailed for six months. Many of them were ill-treated while in detention. They fought their cases, however, and the courts

---

[30] The school itself had been founded in 1951 by Bishnumaya Majhiya and had received financial support, among others from Suyra Bahadur Thapa in the mid-1960s.

eventually upheld their claims that these had been 'forced loans' rather than theft.

The Central Hills

In January 1975, Narayan Man Bijukche (known as 'Comrade Rohit') formed the Nepal Workers and Peasants Party (NMKP), a communist party with an explicit commitment to the peasantry, which has retained a strong base in Bhaktapur. It was initially a break-away from the Communist Party of Nepal led by Pushpa Lal, and an amalgamation of the Proletarian Revolutionary Organisation, Nepal and the *Kisan Samiti* (which may have been linked with the earlier *Akhil Nepal Kishan Sangh*).

The Communist Party Divided

As we have already seen, following the demise of the Jhapa uprising, M B Singh, together with other Maoists including Nirmal Lama, had formed the Central Nucleus of the Communist Party which, following a Fourth Party Congress or Convention in September 1974, became known as the CPN (Fourth Congress or Convention). Its strategy was to launch a people's movement of peasants and workers in the rural areas that could at the right moment be converted into an armed revolt.

The death of Mao Tse Tung and the overthrow of the 'Gang of Four' in China led to intense ideological struggle within the communist movement, particularly among the Maoists. In the CPN Fourth Congress, Nirmal Lama recognized the new Chinese leadership, but M.B. Singh denounced 'Chinese revisionism', labelled the Chinese leadership under Deng Hsiao Ping as 'counter-revolutionaries', and pledged allegiance to orthodox Maoism and the Cultural Revolution. The faction led by Nirma Lal broke with Singh's faction in 1983-84, but continued to operate – albeit now as a separate party – under the name of the CPN (Fourth Congress).

Singh and others formed the CPN (Masal). This divided two years later into two groups, confusingingly called CPN (Masal) and CPN (Mashal) – the former led by M. B. Singh and Babu Ram Bhattarai, and the latter by Mohan Baidya (later known as 'Kiran'). It was in the CPN (Mashal) that Pushpa Kamal Dahal (later known as 'Prachanda') became a leading figure and general secretary in 1989. In 1990, all of these groups, except for the CPN (Masal) under M. B. Singh, split finally from the CPN (Fourth Congress and united under the leadership of 'Prachanda' as the CPN (Unity Centre).

In May 1985, the Nepali Congress Party, which through the 1970s had been involved in an armed struggle against the *Panchayat* System that involved mainly sporadic bombings in urban areas, particularly in the eastern *Tarai* (but also an armed insurgency in Solukhumbu District in December 1974), launched a campaign of civil disobedience (*satyagraha*), starting with a general strike and a 'fill the jails' campaign. The movement appeared to be gaining ground when a series of bombings took place in June 1986, initiated by a group called the Nepal *Janabadi Morcha* (the People's Front), led by Ram Raja Prasad Singh, which aimed to overthrow the *Panchayat* System, topple the monarchy, install a democratic republic and abolish private property.

The Warrior Revolutionary & 'Che' Guevara

Ram Raja Prasad Singh, a political activist whom C. K. Lal has described as "the warrior revolutionary" (Lal, 2013) was born in 1936 in Saptari district in the eastern *Tarai*. Ram Raja recalls one particular incident from his childhood: during a rally organized by the Quit India Movement just across the border from his home, he witnessed a shooting. As the protester died, some of the other dissidents applied his blood to their foreheads and began to shout, 'He has attained martyrdom; he is a martyr'. The incident remained vivid in Singh's memory many years later: 'I realized how sacrosanct was the blood of a revolutionary'.

His father sent him to Banaras Hindu University (BHU), where he spent four years studying English literature and Liberal Arts; but after joining a general strike he got himself expelled. He tried other schools but eventually ended up at the University of Delhi, where he mixed with many different kinds of radicals and effectively began his revolutionary career. In his memoir, Ram Raja recalls his friendship with a South American girl named Clara, who, in 1961 arranged for him and some of his friends to meet Ernesto 'Che' Guevara on an island off the Burmese coast. C. K. Lal comments that, 'nearly half a century after that meeting, Singh still became excited at the mere memory of Che's charisma' (Lal, 2013, p.84). Lal tells us that 'Singh translated some of Che's essays into Nepali, although none of them can now be traced'.

After finishing his eductation in Delhi, Ram Raja studied law at Bihar University and then returned to Nepal in 1964. Three years later, he contested one of the four seats reserved for national graduates in the assembly elections. He argued that 'political transformation in Nepal had to undergo the process of representation, persuasion, agitation, and revolution'. 'Revolutionaries', he declared, should be prepared for 'the annihilation of the old order'. Given such an apocalyptic vision, he was unsurprisingly arrested, barred from the election, and charged with sedition. He secured his release through his familial connections to the Indian establishment and by persuading the court that he had only discussed the theoretical stages of revolution and was therefore not guilty.

By the 1970s, Ram Raja and his party, the Nepali Congress Party (NCP), had begun to disagree about the best strategy for building democracy in Nepal. The Party leadership had lost faith in armed struggle and had handed over their weapons to insurrectionists in East Pakistan. In contrast, Ram Raja, along with many others in the NCP and the Nepali communist movement, thought that reform from within was impossible. Some of these communists began to take up arms (as we have seen), but Ram Raja decided to try a different approach. Once again, he ran as a graduate constituency candidate for the national assembly but promoting a platform that called for an end to the undemocratic *panchayat* system, and warning that if this did not come about, armed struggle might be the only path.

On October 22, 1971, the *Naya Sandeh* weekly reported on a speech Ram Raja made in Biratnagar, writing that the candidate 'went beyond the limits' and told the audience, 'if the *panchayat* system is not abolished peacefully . . . the leaves of the trees will turn into bombs, grains of paddy will become bullets'. He won his seat, marking a significant defeat for the regime, but the state prevented him from taking an oath of office and arrested him. After a special tribunal convicted him, King Mahendra invited him to the palace, a method the monarchy used to woo dissenters, granted him a royal pardon and allowed him to take his place in the assembly.

In 1975, he again toured the country, this time in his official capacity as a NCP activist and an assembly member. His rousing speeches decried the regime's undemocratic control over the nation. Once again, he was expelled from the national assembly and arrested. Indeed, he spent much of his time over the next few years in and out of jail. In 1976, he established the Nepal Democratic Front as a left-wing political movement; the party changed its name to the Multi-Party Democratic Front when King Birendra called a referendum to decide the future of the *panchayat* system in 1980.

To prepare for the vote, other prominent NCP activists were released from jail. The whole spectrum of communist parties — some of which, like the CPN (Fourth Convention) or CPN (Masal) under M.B. Singh, refused to endorse the referendum but were prepared to participate in it. Ram Raja himself began to call not only for multi-party democracy but also for a constituent assembly; in the meanwhile he began to prepare for an armed struggle. He began recruiting fighters from the NCP's left wing, while seeking out weapons and support further afield. He tried to establish training camps in his old stomping ground of Bihar and reached out to the Tamil freedom fighters and Tiger Siddiqui, the nephew of the assassinated Bangladeshi Prime Minister Mujibur Rahman.

When this proved unsuccessful, he went to Chambal, a remote forested region of Uttar Pradesh, and bought guns from the bandits who operated there. One of them recommended that he use explosives rather than guns, and he took this advice. He returned to Nepal and recruited some ex-servicemen from the Nepal army to train his young guerrillas. C. K. Lal remarks that 'not much is known about what Singh and his band of guerrillas did between 1980 and 1985 — he claims they were being trained in handling fire-arms and explosives in different locations around northern India'. Whatever they were doing, they felt sufficiently prepared to act in the summer of 1985. When they did, they intervened not in the rural areas but in the towns, not with guns but with bombs.

That May, the NCP had called for a civil disobedience action (*satyagraha*), which won support from other left-wing activists, including most of the communist parties. The media described the event as a sign of a gathering storm. Then, on June 20, 1985, a series of coordinated blasts in the capital and in other cities rocked Nepal. At least, eight people died, including a member of parliament. In Kathmandu, explosions went off near the royal palace, at the Hotel de l'Annapurna, at the prime minister's office, and near the national assembly, where a politician was killed. A staff member at the Annapurna Hotel also lost her life. In Pokhara, the person responsible for the bomb was killed, and another woman died in Birgunj. Bombs also went off at the Bhairahawa airport, Nepalganj and Mahendranagar in the west, and Janakpur, Biratnagar, and Jhapa in the east.

Singh and his guerrillas admitted responsibility but defended their actions. They explained that they had planned the bombing at the national assembly for a holiday, believing no one would be there. Unfortunately, the rain drove some people to take shelter, turning them into unintended victims. Likewise, the guerillas had left the explosive device on the Annapurna Hotel's lawn, but someone had inadvertently deposited it in the lobby. After the explosions, hundreds were arrested and the Destructive Crimes (Special Control and Punishment Act) was rushed through Parliament. Ram Raja Singh was charged and found guilty in absentia. He, his brother Laxman, Bisheshwar Mandal, and Prem Bahadur Vishwakarma were given the death penalty; others received life sentences, and many had their property confiscated. These named leaders went underground, but five others disappeared while in custody.

After the first spectacular attacks, Ram Raja's operations in Nepal could not continue. He was subsequently arrested in India, but was later released him secretly and allowed to live quietly in Patna. His guerrilla war ended, almost without trace. After G. P. Koirala became the first elected prime minister of the new multi-party democratic order in 1990, however, the state withdrew all charges against him. He even stood as an NCP candidate in the mid-term elections of 1994, but he lost badly and retired finally to live the rest of his days in India. It is not clear what happened to the others involved in his short-lived guerrilla war.

The Jana Andolan

Now, during the late 1980s, all of the left political groups were involved in the mobilisation of various sections of Nepali society in opposition to the *Panchayat* System and the *status quo*, including peasants and agricultural workers. Finally, in 1989, the Ma-Le leadership considered the time ripe to launch another major political thrust. They convened a Fourth Convention and some 90 delegates met under tight security in August in the forest in Udaypur District, near Siraha. They decided to call on all of the 11 communist party factions and groupings, and the Nepali Congress Party, to form an alliance to bring an end to the *Panchayat* System.

By early 1990, an alliance of seven communist parties had been created, as a United Left Front, to constitute – together with Congress – a broader Movement for the Restoration of Multi-Party Democracy. This included, among others, the Workers and Peasants Party of Comrade Rohit. During February and March 1990, support for the democracy movement or the National People's Movement (*Jana Andolan*) grew to the point where it could no longer be ignored. There were mass demonstrations in most of the major cities and all across the country. Peasants and agricultural workers were also involved, but this was a good deal more than a peasant movement – it was a People's Movement - and in April 1990, the *Panchayat* System effectively came to an end. It seemed, for a time, that a new multi-party democracy was being born, albeit with a constitutional monarchy still in place, and that a progressive transformation of the political economy of Nepal was immanent.

This eventually proved not to be the case, and in 1996, the CPN (Maoist) launched a People's War, which gradually gained momentum through the latter part of the decade, as peasants, among others joined the insurgency, and a major 'peasant war' – one of the last of the 'peasant wars of the 20th century' – promised to bring about significant change, for better or for worse. In the third part of this anthropological-historical analysis of rural unrest in Nepal, we intend to consider the period leading up to, during and immediately after the People's War in Nepal.

References

Adhikari, J. (2008). *Land reform in Nepal: Problems and prospects.* Kathmandu: ActionAid Nepal.

Adhikari, S. (1998). *Jumla rajyako itihas.* Kirtipur: Tribhuvan University.

Aziz, B. N. (2001). *Heir to a silent song: Two rebel women of Nepal .* Kathmandu: Centre for Nepal and Asian Studies (CNAS), Tribhuvan University.

Aziz, B. N. (1993). *Shakti Yogmaya: A tradition of dissent in Nepal.* Kathmandu: Centre for Nepal and Asian Studies, Tribhuvan University.

Bailey, A. (1974). *On the specificity of the Asiatic mode of production.* M. Phil. thesis, University of London.

Basnet, J. et al. (2010). *Asserting freedom from central control,* Nepal Country Paper, Land Watch Asia, Securing the Right to Land. Kathmandu: Action Aid International, International Land Coalition & MISEREOR.

Basnet, J. (2008). Overview of land rights movements in Nepal. In B. Upreti,S.R. Sharma and J. Basnet (Eds.) *Land Politics and Conflict in Nepal.* Kathmandu.

Bishop, B. (1990). *Karnali under stress: Livelihood strategies and seasonal rhythms in a changing Nepal Himalaya.* Chicago: University of Chicago Press.

Blaikie, P., Cameron, J., & Seddon D. (1980, 2001) *Nepal in crisis: Growth and stagnation at the periphery.* Delhi: Oxford University Press.

Blaikie, P., Cameron, J., & Seddon D.(1979). *The struggle for basic needs in Nepal.* Warminster: Aris & Phillips.

Budhathoki, B.(2048 vs.). *Nepalma kishan andolan (The Peasant Movement in Nepal),* Bhaktapur.

Cailmail, B. (2008-09). A history of Nepalese Maoism since its foundation by Mohan Bikram Singh. *European Bulletin of Himalayan Research,* no.s 33-34, Autumn 2008-Spring 2009, Pp. 11-38.

Cailmail, B. (2013). The CPN (Masal) Bastion: A history of communist militancy in Pyuthan District and its influence on the people's war. In M. Lecomte-Tilouine(Ed.)*Revolution in Nepal: An anthropological and historical approach to the people's war.* Delhi: Oxford University Press.

Caplan, L. (1970). *Land and social change in east Nepal: A study of Hindu-tribal relations.* London: Routledge & Kegan Paul.

CERM. (1971). *Sur le Féodalisme,* Centre d'Etudes et de Recherches Marxistes. Paris: Editions Sociales.

CERM. (1970). *Sur les societiés précapitalistes,* Centre d'Etudes et de Recherches Marxistes. Paris: Editions Sociales.

CERM. (1969). *Sur le 'Mode de Production Asiatique'.* Centre d'Etudes et de Recherches Marxistes. Paris: Editions Sociales).

Conway, D., & Shrestha N. R. (1981). *Causes and consequences of rural to rural migration in Nepal.* Indiana: Indiana University Press.

Cox, T. (1990). Land rights and ethnic conflicts in Nepal. *Economic & Political Vernacular Nepali Weekly,* 16 June 1990, Pp. 1318-1320.

Communist Party of Nepal (CPN) (1949) *Nepal communist party ko pahilo ghosana patra (The First Manifesto of the Communist Party of Nepal),* Kathmandu: CPN

CSRC. (2006). *Haliya: Bhumi adhikar bare padhne samagri* (Haliya: Reading Material on Land Rights). Bhumi Adhikar (Land Rights), 3:2. Kathmandu: Community Self Reliance Centre.

Dhakal, S. (2013). *Land and agrarian questions: Essays on land tenure, agrarian relations and peasant movements in Nepal.* Kathmandu: Community Self-Reliance Centre.

Dhakal, S. (2013). Chintang Bidorha 1979: Locating a peasant movement in a political landscape. In S. Dhakal. *Land and agrarian questions.* chapter 6. Kathmandu: Community Self-Reliance Centre.

Feldman, D. & Fournier, A. (1976). Social relations and agricultural production in Nepal's *Tarai. The Journal of Peasant Studies,* vol.3, no.4, Pp. 447-464.

Fitzpatrick, I. C. (2011). *Cardamom and class: A Limbu village and its extensions in east Nepal.* Kathmandu: Vajra Publications.

Gallissot, R. (1975). Pre-colonial Algeria. *Economy & Society,* vol. 4, no. 4.

Gellner, E. (1969). *Saints of the atlas.* London: Weidenfeld & Nicolson.

Ghimire, K. (1992) *Forest or farm? The politics of poverty and land hunger in Nepal.* Delhi: Oxford University Press.

Hobsbawm, E. (1964). 'Introduction' to K. Marx, *Pre-capitalist economic formations.* London: Lawrence & Wishart.

Hobsbawm, E. (1959). *Social bandits and primitive rebels: Studies in archaic forms of social movements in the 19th and 20th centuries.* Glencoe: The Free Press.

INSEC. (1992). Bonded labour in Nepal under kamaya system. Kathmandu: INSEC.

Kaplan, P., & Shrestha N. R. (1982) The sukumbasi movement in Nepal: The fire from below. *The Journal of Contemporary Asia,* vol. 12.

Karki, A. (2002) *The politics of poverty and movements from below.* Unpubl. PhD thesis, University of East Anglia.

Karki, Arjun. (2002). Movements from Below: land rights movements in Nepal. *Inter-Asia Cultural Studies,* vol.3, no 2, August.

Karki, A. & Seddon, D. (2003). *The people's war in Nepal: Left perspective.* Delhi: Adroit Publishers.

Kishan, S. (1977). *An official bulletin of the Nepal mazdoor kishan party*, no.3, Bhaktapur.

Lacoste, Y. (1974). General characteristics and fundamental structures of mediaeval North African society. (translated by David Seddon). *Economy & Society*, vol.3, no.1.

Lal, C. K. (2013). Ram Raja Prasad Singh: The warrior revolutionary. In M. Lecomte-Tilouine (Ed.) *Revolution in Nepal: An anthropological and historical approach to the people's war*. Oxford: Oxford University Press.

Lecomte-Tilouine, M. (2013). *Revolution in Nepal: An anthropological and historical approach to the people's war*. Oxford: Oxford University Press.

Lecomte-Tilouine, M. (2003). The history of the messianic and rebel king Lakhan Thapa: Utopia and ideology among the *Magars*. In D. Gellner (Ed.) *Resistance and the state: Nepalese experiences*. New Delhi: Social Science Press, Pp. 244-78.

Lecomte-Tilouine, M. (2000). Utopia and ideology among the *Magars*: Lakhan Thapa versus Mao Dzedong? *European Bulletin of Himalayan Research*, 19, 73-97.

Manandhar, P. & Seddon, D. (2010). *In hope and in fear: Living through the people's war in Nepal*. Delhi: Adroit Publishers.

Mikesell, S. (1999). *Class, state and struggle in Nepal*. New Delhi: Manohar.

Murray, R. (1975). Class, state and the world economy: A case study of Ethiopia. Paper Presented to the Conference on 'New Approaches to Trade', IDS, Sussex, September 1975.

Neupane, D. (2015). *The synthesis between social protest and nirguna bhakti in Yogmaya's sarvartha yogbani*. Kathmandu: Tribhuvan University.

Neupane, D. (2012). Spiritualism and religious fervor in sarvartha yogbani. *Pursuits: A Journal of English Studies*, July 2012.

NRSU. (2011). The role of Nepal workers and peasants' party in the communist movement of Nepal. *The Workers' Bulletin*, 1.1 (1): 1–6.

Pandey, B. & Rimal, B. (translation by Mukunda Kattel). (n.d.). *Women participation in Nepali labour movement*. Kathmandu: GEFONT.

Pokharel, C., & Basyal, R. (1999). *Rapti paschim ky prajatantric andolan (The democratic movement in west Rapti)*. Bardiya: Subash Pokhrel.

Pradhan, K. (1991, 2009). *The Gorkha conquests: The process and consequences of the unification of Nepal with particular reference to eastern Nepal*. Calcutta: Oxford University Press & Kathmandu: Social Science Baha/Himal Books.

Raj, Y.(2010). *History and mindscapes: A memory of the peasants movements of Nepal*. Kathmandu: Martin Chautari.

Rana, P. S. J. B., Rana, P. S. J. B., & Rana, G. S. J. B. (2002). *The Ranas of Nepal.* Geneva: Edition Naef. Kister S.S. Editeur.

Rana, P.S. (2000). *A chronicle of Rana rule.* Kathmandu: R. Rana.

Rana, P.S. (1995). *Rana intrigues.* Kathmandu: R. Rana.

Rana, P.S. (1978). *Rana Nepal : An insider's view.* Kathmandu: R. Rana.

Rawal, B. (2007). The Communist movement in Nepal: Origin and development. Kathmandu/Accham: Forum CPN (UML).

Regmi, M. C. (1999). *A study of Nepali economic history, 1768-1845.* New Delhi: Manjusri Publishing House; Delhi: Adroit Publishers.

Regmi, M. C. (1988.) *An economic history of Nepal: 1846-1901.* Varanasi: Nath Publishing House.

Regmi, M. C. (1984). Bhote and Murmi rebels. *Regmi Research Series,* 16: 129-30.

Regmi, M. C. (1977). *Land ownership in Nepal.* Delhi: Adroit Publishers.

Regmi, M. C. (1963-68). *Land tenure and taxation in Nepal* (4 vols). Berkeley: Berkeley Institute of International Studies, University of California.

Sah, R. M. (2017). *The middle country: The traverse of Madhes through war, colonization and aid dependent racist state.* Delhi: Adroit Publishers.

Scott, J. (1985). *Weapons of the weak: Everyday forms of peasant resistance.* Yale University Press.

Seddon, D. (1981). *Moroccan peasants.* Folkestone, Kent: Wm Dawson.

Seddon, D., Blaikie, P., & Cameron, J. (1979, 2001). *Peasants and Workers in Nepal,* Delhi: Vikas Publishers.

Shrestha, P. (2032 VS, 1977). *The manifesto of the all Nepal peasant association (ANPA) and its strategies.*

Shrestha-Schipper, S. (2013). The political context and the influence of the people's war in Jumla, In M. Lecomte-Tilouine (Ed.). *Revolution in Nepal: An anthropological and historical approach to the people's war.* Delhi: Oxford University Press.

Singh, M. B. (1992). *Char dashak: Char dashak ko bibadha, birodha ra sangarsha ko rajniti (Four Decades: Four decades of controversy, opposition and struggle).* Gorakhpur: Janashahitya Prakashan

Terray, E. (1974). Long distance trade and the formation of the state: The case of the Abron kingdoms of Gyaman, *Economy & Society,* vol.3, no. 3.

Thapa, S. (2001). *Peasant insurgence in Nepal: 1951-1960.* Bhaktpur: Nirmala KC.

Thapa, S. (2000). *Historical study of agrarian relation in Nepal (1846-1951).* Delhi: Adroit Publishers.

Thapa, S. (1996). An approach to the peasant movements in Nepal, 1951-1960. *Tribhuvan University Journal,* xix, 73-88.

Thapa, S. (1995). *Agrarian relations in Nepal: Owners-cultivators in a traditional agriculture.* Unpubl. Ph.D Dissertation, Patna University.

Upreti, B., Sharma, S. R., & Basnet, J. (Eds). (2008). *Land, politics and conflict in Nepal: Realities and potentials for agrarian transformation.* Kathmandu: Community Self-Reliance Centre, South Asia Regional Co-ordination Office of NCCR North-South, and Human and Natural Resources Studies Centre (HNRSC), Kathmandu University.

Walton, J. & Seddon, D. (1994). *Free markets and food riots: The politics of global adjustment.* Oxford: Blackwells.

Whelpton, J. (2013). Political violence in Nepal from unification to *Jana Andolan* I: The background to the people's war. In M. Lecomte-Tilouine (Ed.) *Revolution in Nepal: An anthropological and historical approach to the people's war.* Oxford: Oxford University Press.

Wolf, E. (1971). *Peasant wars of the twentieth century,* London: Faber & Faber.

The Landlord State, Adivasi People, and the 'Escape Agriculture' in the Eastern Tarai of Nepal: A Historical Analysis of the Transformation of Dhimals into a Farming Community

JANAK RAI

Introduction: "Our ancestors were less interested in adding land"

"Even when people from elsewhere started to settle in our areas in the late 1950s, and began encroaching and taking the available land by clearing the forest, our ancestors were less interested in adding land for their families. They still enjoyed the forest-based ways of life and used to spend many months hunting in the forest. Some of our ancestors did have some wealth--silver, some gold and coin money. Rather than buying additional lands with this wealth when land was plentiful and cheaper, our ancestors would keep these possessions in an earthen jar and bury under the soil" Mr Kabin (name changed), a 56 year Dhimal man from Belbari area of Morang, told me in 2008. We two were discussing about the problem of landlessness among the *Dhimals*, an indigenous community from the easternmost *Tarai* of Nepal. At that time, Mr. Kabin was a member of the central executive committee of the *Dhimal Jati Bikas Kendra* (hereafter the *Kendra*), the national level indigenous organization of the *Dhimal* community. A well-respected "indigenous intellectual" (Rappaport, 2004) and an active indigenous right activist affiliated with the *Kendra*, Mr. Kabin is a very knowledgeable person on *Dhimal* culture and history. In the above-mentioned excerpt of a long taped interview, he was explaining about how *Dhimal* ancestors, as late as 1950s, *had little interest in accumulation of land* for future when people from the hills and elsewhere were settling in Morang and Jhapa because of the easy availability of agricultural lands. Like many other research participants with whom I discussed about the question of landlessness, Kabin also underlined that until the recent past (early 20th century) *Dhimals* had different understanding about and relationship with land (see Rai, 2013a). They were not 'peasants' in the sense of a community which has "cultivation of land and animal rearing as the main means of livelihood" (Shanin, 1973).

During my fieldwork (2007-2009) in Morang, many senior *Dhimal* informants (sixty years and above) told me that their ancestors preferred not to own land even when land was plentiful in the *Tarai*. "Owning land was so much of *dukkha* (hardship, pain) in the past," they would say (see Rai, 2015). Instead of owning land under their names, many villagers were said to have opted to work as tenants in the landed *Dhimal* families. I used to find these explanations to be puzzling and even contradictory in the similar way the anthropologist Arjun Guneratne (1996) had experienced in his study of the Chitwan *Tharu*[1]. The *Tharu* villagers would often tell Guneratne that in the "old days" (before 1960s) – "land was plentiful and to be had for asking. But.....that many landless and near landless families in the village, as long as anyone could remember, used to earn their livelihood by working as servants for their landed neighbors" (Guneratne, 1996, p.5). If the *Tharu* and *Dhimal* were 'peasants', why would they not own land even when it was readily available? Such non-owning of the land or the practice of "volunteer landlessness" (Guneratne,

---

[1] I had not read Arjun Guneratne's (1996) article at the time of my fieldwork. Guneratne does not bring this issue in his book (Guneratne, 2002) which I had read before.

1996, p.31) provides a striking exception to "the consensus among students of peasant societies is that the control of land is the most important and desirable guarantee of subsistence in the peasant economy" (p. 5).

Scholars generally approach the 19th - early 20th century agrarian societies in Nepal by privileging *the state-peasant relations* and thus ignore the historical reality of non-farming communities, particularly the *Tarai adivasis* such as the *Dhimals*. In this paper, I draw on the *Dhimals' historical analysis* to discuss why certain sections of the *Tarai adivasis* preferred not to 'own' land in the past. The case of *Dhimals*, as this paper will show, asks us to rethink the relevance of the concept of 'peasantry' as an analytical category in examining the changing relations between the Nepali state and the *Tarai adivasis* over the control of land in the past.

It was the anthropologist Lionel Caplan (1990, 1991) who emphasized the usefulness of the distinction between 'tribal' (*adivasi*) and 'peasant societies' as analytical categories for Nepal. His distinction between these two societies is based on their different understanding and relationship to the land. This paper also demonstrates the relevance of the distinction of 'adivasi' and 'peasant' as analytical categories in examining the history of changing relationship between the landlord state-the *Tarai adivasi* and understading the transformations of the *adivasi* into peasant societies during the 19th and early 20th centuries. I also discuss *Dhimal's* explainations on how their traditional village head or *Majhi* emerged as local landlord in the early 20th century. My analysis highlights how people's strategy of evading the oppressive and extractive landlord state led to the emergence of distinct class relations based on land ownership among the *Dhimals* in the early 20th century. Informed by the historical studies such as that of Mahesh Chandra Regmi (1987 and 1979), I use the 'indigenous mode of historical analysis' (Kirsch, 2006) offered by the *Dhimals* themselves to discuss why their ancestors preferred not to enter into the tenurial relationship with the state. I argue that in order to understand how the the majority of the *Dhimal* became 'landless' after 1970s, we need to pay closer attention to that the specific ecological and political conditions under which the *Dhimal* relationship with the land altered when they experienced the landlord state in the *Tarai* during the 19th and early 20th centuries.

Adivasi and Peasantry: Conceptual Discussions

In this paper, I have used the concepts of *'adivasi'* and *'*peasant*'* as analytical categories to emphasize distinct societies with respect to their relationships to the land, nature of economic system, and relationships with the state over the control of land. If we consider *'adivasi* and *'*peasant*'* as analytical categories, then the centrality of *'*land*'* in defining the two categories become obvious. *'*Land*'* is an important material possession (a physical entity, property, a resource) with its cultural and symbolic relevance for peasants. For the *adivasi*, land is a defining feature of their collective identity but it is primarily an inalienable wealth, not a commodity. The peasant-land relationship draws heavily on capitalist ontology of commodity and property while the *adivasi*-land relationships emphasizes the total embeddedness and mutual production of land, people and culture in totality.

The term *'*peasant*'* is generally used to describe *'*rural*'* households with rights to land in cultivating primarily for their own subsistence and not for market (Shanin, 1971). But this simple definition, which focused on rights to land and cultivation of land primarily for subsistence, does not address the political-economic relations of such production with the dominant classes. In his classic study, Eric Wolf (1966) distinguishes between *'*peasants*'* and "primitive people"[2]. Here, I use *'*tribal people*'* to avoid negative connotations associated with the concept of "primitive people". According to Wolf, the tribal people "live in countryside and raise crops and livestock" (p.2) and "are involved in the wider network of reciprocal exchange relations with other groups (p.3)". For Wolf, the distinctive feature of such *'*tribal society*'*, which distinguishes it from *'*peasant society*'*, is that the producers in the former society "control the means of production, including their own labor, and exchange their own

---

[2] Wolf uses the concept of "primitive people" to refer to the tribal people without hesitation of the colonial baggage and derogative implications we know associate with the term "primitive" in anthropology.

labor and its products for the culturally defined equivalent goods and services of the others" (p.3). But peasants, however, are "rural cultivators whose surpluses are transferred to a dominant group of rulers that uses the surpluses to underwrite its own standard of living and to distribute the remainder to groups in society that do not farm but must be fed for the specific goods and services in turn" (Wolf, 1966, p.4). In other words, tribal people are more autonomous in terms of sharing of their products and labor among themselves while peasant producers not only produce for themselves but also for other dominant groups who appropriate their surpluses. Wolf's emphasis on the political-economic relationship and control of means of production in making distinction between 'tribal' or 'indigenous' producers and peasant producers are relevant for this paper.

In the present context, the fundamental feature of ādivāsi identity is considered to be the *inalienable relationship* they have with their land and territories (Gray, 1995, Castree, 2004). Using the case of the *Limbu* and their inalienable relationship with the *Kipat* land, the anthropologist Lionel Caplan (1990) has persuasively argued that ethnic groups like the *Limbu* can be considered 'tribe' (read *adivasi*) "in terms of their traditional relationships to land, and how this attitudes to land shaped people's identities" (p.135). The *Limbus* are distinctively 'tribe' (indigenous) because they do not consider land as a commodity. For them, *Kipat* stood for their way of life, thus symbolized cultural vitality, and continuity of the community (Caplan, 1990, p.137). In contrast to this, the non-tribal peasants relate to land in a different way; the *raikar* land became a private property by the later part of the 19th century (p.138) and hence, had transformed into an alienable commodity (p.140)[3]. Caplan's argument that the distinction between 'tribe' and 'peasant' at analytical level can help us to understand the transformations of the former into later is insightful and relevant for the case of Dhimal as well. In other words, the

---

[3] It should be emphasized here that Caplan (neither I) do not suggest that peasants do not have emotional, symbolic or non-economical relationships with the land. However, for peasants (at the level of its definitional level), their relationship with land is defined in terms of property ownership – a material possession out of which they make their survival.

distinction of 'tribe' and 'peasant' in terms of their relationship to the land is useful to understand the different relationships these two groups with the Nepali state in the past.

In this paper, I have attempted to discuss the concepts of *'adivasi'* and 'peasants' from the historical perspectives of *a non-farming community.* James Scott's (2009) works on the state repelling practices in the highland of Southeast Asia can be insightful in approaching the historical choice of not owning land by the *Tarai adivasi.* In his book, *The Art of Not Being Governed* (2009), Scott draws on various state evading practices adopted by many upland communities in Southeast Asia to argue that people can deploy physical mobility, subsistence practices, social organizations, and settlement pattern, often in combination, to place distance between a community and state appropriation (pp.182-183). According to Scott (2009, pp.178-219) subsistence practices such as foraging for forest foods, adoption of shifting cultivation, and rejecting the settled agriculture practiced by many upland communities in the Southeast Asia were a historical choice and political strategy, what he calls "state repelling" techniques (p.180), adopted by upland groups to escape state appropriation. Given the extractive nature of the 19th century Gorkhali state and its oppressive regimes of revenue collection (see Regmi, 1971, 1978 in particular), I used the case of the *Dhimals* (Rai, 2015) to argue that the practice of not owning the land also needs to be seen as a political choice or what James Scott (2009) has called "escape agriculture" in order to evade the state power.

Scott's analysis is insightful, however his concept of "Zomia" is less applicable to describe the *Tarai ādivāsi. Dhimal* were not the "Zomia" population in the ways in which Scott uses the concept to describe the Southeast Asian highlanders who resisted slavery, conscription, taxes, *corvee* labor, warfare, and complete assimilation into the state governance by fleeing the oppressions of state-making projects in the valley and lowlands. Similarly, his sweeping characterization of the Zomia as "runway," "fugitive" and "stateless" people can potentially misrepresent the ways in which indigenous peoples exercised their political autonomy (including territorial autonomy), and resisted the Hindu landlord state in

Nepal (Dambar Chemjong, 2010, personal communication). Similarly, elsewhere I have argued how Scott's (2009) analysis is inadequate to understanding the overarching moral and political rationalities that Nepal rulers' envisioned and used to legitimatize their sovereignty over the people and territories they claimed to govern (Rai, 2013a). Despite these limitations, Scott's concept of 'escape agriculture' as a state repelling technique is useful to examine how people could adopt non-farming practices in order to subvert the extrative state power. My intention in this paper is to use *Dhimal's* historical experience and analysis to argue that 'peasantry' should not be taken as the generalizing deductive framework to explain the state-a*divasi* relationships over the control of land during the 19th-20th centuries in Nepal. It is relevant here to underline a major critique against the earlier subaltern scholars such as Ranjit Guha for ignoring the distinction of *adivasi* communities and "subsuming all tribal revolts under the rubric of peasant insurgency in his work on peasant consciousness" (Kela,2006, p.506). Hence, the distinction between *adivasi* and peasant communities become imporant for understanding the specific political history over the control of land in the easternmost *Tarai* from the perspectives of non-farmimg *adivasi* who thrirved along the fringes of the densed malarial forests.

The Ecology of Belonging and Different Understanding of 'Land'

Dhimal were living in the easternmost lowlands long before the rise of the present day Nepali state. Until the early twentieth century, most of the plains of Morang were thinly populated and thickly forested. At that time, Morang, which also included the present districts of Sunsari and Jhapa, was considered to be "extremely swampy with its pestilent climate...the most malarious and unhealthy district" (Oldfield 1881, pp. 61-622).

Hence, these malarial lowlands were also called *Kala Pani,* literally meaning 'black water', a 'deadly place' where outsiders would definitely die if they stay for long. While outsiders feared the region, the aboriginal inhabitants such as *Dhimal, Meche, Tharu, Koch* and others survived the malarial environment and transformed the seemingly inhabitable places into their homelands. Then, *Dhimals* lived more of a *migratory village life* following an ecological niche that availed them plentiful resources: fish, wild animal and plants to survive without much competition with other human beings.

The present day *Dhimals* emphasized that farming under such forested ecology was neither practical nor desirable, and hence, their ancestors followed forest-based ways of life. *Dhimal* ancestors subsisted by foraging, hunting, fishing, periodic farming, and engaging in exchange relationships with the neighboring groups from the hills and India. In their retellings of the past, my *Dhimal* friends invariably mentioned the threats of *wild animals, malaria* and *cholera* to emphasize the everyday challenges their ancestors encountered when they transformed the dense forested *Tarai* into their dwelling places. People were forced to move from one place to another due to the outbreaks of *haija* and threats of wild animals. The low yields of the harvest and the destruction of crops by wild animals made farming less practical, and hence, people could not rely on farming to sustain their lives (see Rai, 2013a for how Dhimal narrates the malaria past). Hence, the concept of 'peasant' is less useful to describe how this specific history of ecological belonging was in the *Tarai* of Morang until the early 20th century.

Since *Dhimal* ancestors lived under such ecological conditions until the late 19th century, their relationship and reliance on the land was very different from the ways in which the rulers related to it *as state property rentable* to the subjects with certain obligations. In the first scholarly account of the *Dhimals* (in India) published in 1847 by Brian Hodgson (1880 [1847]) described the *Dhimal* people as "erratic cultivators of the wild (p.117)" and that they "*claim no proprietary or possessory ownership…*" (p.119, emphasis added). Hodgson's observations that *Dhimals* during the mid-19th century had no conception of land as a proprietary possession suggests that their sense of territorial belonging at that time was not strongly based on cultivation of crops but more on the components of their ecological niche -- earth/soil, forests, rivers, animals, sacred places, and others.

It is difficult to ascertain how *Dhimal* ancestors understood land in the 19th century; we do not have definitive data about it. However, the *Dhimali* terms for 'soil' and 'land', and their core ritual symbolisms can help us to infer how their ancestors may have understood land when they lived in the forested and malarial habitats in the Tarai in the past. Now Dhimal make distinction between *bhonai* and *meeling* when they refer to 'land'. *Meeling* is more strictly used to refer to land in the sense of its proprietorial possession and ownership or their lack of it (akin to Nepali word for land '*jaggā*'). On the other hand, the word *bhonai* is used in a broader sense that encompasses the notion of soil or earth. It hews closely to the Nepali/ Sanskrit concept of *bhumi* (earth/soil) and the indigenous articulation of land as inalienable wealth (see for example, Castree, 2004, Caplan, 1970).

I should emphasize here that *Dhimal* religious worldviews and ritual practices are shaped by their historical experience of the challenges of living in the *Tarai* (see Rai, 2013a). A fundamental aspect of *Dhimal* ritual is the recognition of all the agents and material objects that contribute to their sustenance and well-being throughout the year. Forests, rivers, soils, wild animals, and other beings in "nature" are reckoned and honored during their rituals. In other words, the ritual emphasizes the mutual coexistence of various beings based on ethnics of reciprocity. *Dhimals* explain that their ancestors began these ritual practices in order to survive malaria, wild animals, and other possible threats when they were living in the forested ecology in the past.

The concept of *bhonai* as an embedded relationship among humans, deities, and soil suggests that in the past *Dhimal* relationship to the land was mediated through their notion of *bhonai* rather than *meeling*. Unlike the tenurial system of the landlord state that embraced the superiority of the "giver" over the "receiver" of the land, the *Dhimal* cultural ethos of exchange and reciprocity structured their use of the land. The interface between the *adivasi* relationship and understanding of land, and the practical constraints of farming under then *Tarai* environment must have discouraged the *Dhimal* ancestors to rely on farming for subsistence. But beside these the prevalent ecological conditions and indigenous understanding of land, the extractive and oppressive nature of the 19th century landlord state and its hierarchical tenure system further distracted the *Dhimal* ancestors from reclaiming the land for the state and its rulers.

State-led Colonization of Land, Labor and Adivasi

After the Gorkhali conquest, *Dhimal* ancestral territories became, in effect, an internal colony of Kathmandu (Sugden,2009) and the Gorkhali king became the sovereign *malik* (owner, landlord) of the land (see Burghart,1984). Under the state landlordism, the king was the sovereign owner of the land, hence a *bhupati*[4], and the state was the effective landlord[5]. The 19th century tenurial system was "a control hierarchy in which the diverse subjects of his kingdom were brought together by virtue of their tenurial relations to the king" (Burghart,1984,p.112) such that "submission of such payments through tiers of the tributary, civil, and military administrations indicated one's inferiority to the recipient of such payments, and thereby defined the hierarchical structure of the tenurial system" (p.104). In effect, the system "meant that the surplus produce of the land belonged to aristocratic and bureaucratic groups in the society, where the peasant was *a mere instrument to work the land and produce for their benefit* (Regmi, 1978, p.33 emphasis added). In other words, to become a tenant subject in the 19th century Nepal was to accept one's inferior position within this control hierarchy. The state-led economy of the extraction based on the territorial sovereignty of the *bhupati* king was radically different than the ways many *adivasi* communities conceptualized their relationship with land and from their indigenous land ownership practices (Caplan,1970). To use Wolf's distinction between the tribal people and peasant societies (Wolf 1966), the Gorkhali tenurial system undermined the political autonomy and reciprocal relationship of production and exchange relations of the tribal community.

---

[4] See Rai (2013a) for the notion of the king as *Bhupati* and the politics of territorial sovereignty in Nepal .

[5] The actual cultivators usually held land on tenancy and paid rent taxes, rent, and provided free labor service either to the government or to individuals or institutions who were beneficiaries of state land grants (Regmi, 1971).

The noted historian Mahesh C. Regmi (1978) has so brilliantly described the pauperized conditions of the peasantry during the 19th century landlord state in Nepal. Therefore, it is very likely that *Dhimal* ancestors, who had already experienced the oppressive nature of the Gorkhali political order, continued to live along the malarial forests in order to avoid the emerging landlord state while they were also impacted by the emergent socio-political and economic changes in the region. In 1849, Brian Hodgson wrote that the *Dhimal* people whom he had met at the border between the Nepal and Indian *Tarai* regions along the Mechi river informed him that they had come there from Morang 60 years earlier in "order to escape the Górkhali oppression" (Hodgson, 1849, p. 131). Thus, the Gorkhali annexation of the far eastern region, and the territorial disputes and wars with Sikkim, Tibet, and Bhutan had greatly weakened the political and territorial autonomy of indigenous communities such as the *Kirati* (*Rai, Limbus,* and *Sunuwars*) in the hills (Caplan,1970, English, 1983, Pradhan, 1991) as well as in the plains (Gaige, 1975, Guneratne, 2002, Sugden, 2009).

James Scott (2009) argues that as a way of evading state power, people can strategically choose to stay in locations peripheral to state power such as those located in remote and inaccessible for outsiders. And, hence, in such contexts, peripheral locations, Scott asserts, "must be treated as a social choice, not a cultural or ecological given" (p. 183). Hence, what Scott calls "escape agriculture" is not an act of "primitiveness" or "backwardness" often people associate with foragers people, but a conscious choice groups like the *Dhimal* ancestors made in order to blunt the oppressive blow of the landlord state. The thick forests and densely malarial regions of the lowlands of Morang were still peripheral spaces in terms of the direct state control and regulation during the 19th century. The forested and malarial environment provided *Dhimals* with a relative autonomy to evade the oppressive landlord state and continue their traditional ways of life (see Rai, 2013b). Therefore, it is important to highlight here that their preference to continue non-farming ways of life did not simple stem from their habitual dispositions of living in the forested and malarial environment. *Dhimals'* practical experiences of the challenges of cultivating the soil for their subsistence with obligations of the paying tax to the state were far less encouraging but full of *dukkha* (hardship, pain) (see Rai, 2015 for detail).[6] They could still easily and comfortably continue their traditional subsistence patterns when the resources were plentiful. And, most importantly, by not becoming tenant peasants, they could also avoid dealing with the landlord state and those who represented state power in the *Dhimal* territories (Rai, 2015).

The eastern *Tarai* was an important colony for the hill rulers in terms of the revenue it generated through agriculture, sale of forest products, and most importantly through transnational trade between Nepal, British India and other regions. However, considerable areas of land in the lowlands of Morang were still forested, places where cattle herders from neighboring Indian villages would send their animals for pasture during the dry seasons (Buchaman-Hamilton, 1928, pp.414-416). After its conquest, the Gorkhali rulers "consolidated the development of a feudal mode of production in the Morang plains and considerably strengthened the power of the bureaucracy and its capacity to appropriate surplus from Morang" (Sugden, 2009, p.129). The district territory was furthered divided into many revenue collection units called-*Pragana* in which a *chaudhari* (generally a local headman) was appointed to collect revenue (land tax, rent, levies, and fines,) and to promote land reclamation and settlement (Regmi Research Series 1970, pp.107-109). Unclaimed and forested lands were allotted as *birta* land grants to civil and military officials, members of nobility, chieftains of the conquered hill principalities, and other supporters of the ruling regime. Government employees or functionaries were paid for their service through *jagir* land grants in Morang. It was the responsibility of the recipients of these land grants to recruit tenants, including some from India, to expand agriculture in their land holdings.

---

[6] Some of these major challenges *Dhimal* mentioned are: practical difficulties of farming (lack of labor) against the low agriculture yields, the damage of crops by wild animals, lack of market to sell the crops so that they could pay the required land tax in cash, the heavy burden of tax, lack of tenurial security and eviction from the cultivated land in the failure to pay tax, and oppressive and violent nature of the state revenue officials and the landlord (see Rai, 2013 for detail).

As early as 1799-1800, the Nepali state had attempted to promote settlement programs in Morang whereby the settlers could receive "as much land as one could reclaim" (Ojha,1983, p.25). 'Waste land' or unclaimed lands could be freely allocated to any individual from Nepal or India willing to settle and reclaim these lands for farming; tax remissions were made for an initial period that ranged from four to ten years (Regmi, 1971, p.144). However, even with such seemingly liberal state efforts, the resettlement programs failed to attract the desired numbers of people from outside to the *Tarai* of Morang. The fear of malaria was one but not the only factor that had discouraged people from reclaiming land in the Morang region. The political environment -- that of exploitative and oppressive tax regimes in Nepal on the one had, and the prospects of cash-based wage labor in the colonial plantations located in the bordering regions of India, on the other hand -- seemed to have motivated Nepali migrants to choose the malarial *Tarai* of India over Nepal's *Tarai* (Ojha, 1983). Historical records dating back to 1779 show that large numbers of *ryots* (tenant peasants) had left Morang and emigrated to India due to the high level of taxation and oppression of the revenue functionaries (Regmi Research Series 1971, pp.249-251). Hence, labor, not the land, became a major problem for the extractive the 19th century Nepali state in the *Tarai*. And, the shortage of labor together with the absence of hill immigrants created pressure even on non-farming the *Tarai ādivāsi* such as *Dhimals* to reclaim land for the state or to become tenants of absentee landowning elites from the hills. *Dhimals* could not escape the state-led colonization of the *Tarai* land for long.

Increasingly, *Dhimals* had no other options than to rely on cultivation of land for subsistence. Moving out of their villages and resettling in new places was perhaps the last resort, not the alternative *Dhimals* would have preferred in order to ameliorate the brunt of the oppressive landlord state and its representatives. They were less likely to work at wage labor in the colonial planation economy in neighboring districts of India (Hodgson, (1880 [1847], p.119; Waddell 1899, pp. 4-5)[7]. It is very likely that by the

---

[7] Waddell (1899, 4-50) observed (in India) "....the *Mech* and *Dhimal,* who live in the depths of these forests, and *who will undertake no hired service,* have acquired

end of 19th century, *Dhimals* relied more on farming and but continued to follow their customary ways of subsistence pattern, as indicated by Kabin's narratives cited in the beginning of this paper. Land was abundant, and they reclaimed it when needed but they experienced that 'owning' land i.e. by entering into tenurial relationship with the state was full of '*dukkha*'. The practical constraints of farming -- such as the lack of labor, low yields, and threats of wild animals.... always possessed challenges for the *Dhimal* ancestors. On the top of such practical problems, the burden of tax, oppressive and extractive tax collectors and landowning elites, and lack of tenurial security further evicted the *Dhimal* ancestors from their land and ancestral territories (see Rai, 2015).

From the perspectives of the *bhupati* king, the tenants from Dhimal community were seen to be accepting the tenurial sovereignty of the king (the state), and agreeing to become morally and legally accountable to pay the required taxes, rents, and levies and to provide free labor services to the '*malik*' of the land. When *Dhimals* reclaimed land for the state and became its tenants, they were required to produce not only for themselves but also for the state and other "parasitic groups" (Regmi 1978, p. x) who had rights to extract rent and levies because of these groups' tenurial ownership of the land on which *Dhimals* labored so diligently. This was a contradictory political economic relation imposed upon *Dhimals* whereby they, by virtue of reclaiming the land that they had always used, also became a subordinated peasant class subjected to payment of tax and labor services to those who claimed ownership of their ancestral territories. In other words, *Dhimals* first had to own or 'possess' land in order to be exploited and then many would again become landless when they failed to pay land tax. Hence, it was the "dispossession by accumulation" (cf. Harvey, 2004) of land that characterized experiences of *Dhimal* ancestors when they became the tenants of the landlord state in the 19th century[8]. Hence, *Dhimal* ancestors, as long as they could, preferred

---

almost as much immunity from the deadly fevers of these forests....." (emphasis added).

[8] David Harvey (2004) uses the concept of "accumulation by dispossession" to discuss how the neoliberal capitalism works by concentrating accumulation of

not to 'own' land and become tenants for the landlord state. In the next section, I discuss about the emergence of the landlord-tenant relations to highlight the historical processes of the transformation of *Dhimal* into peasant economy.

Emergence of Landed Classes: Dhimal Majhi and the Landlord State

Each *Dhimal* village has its traditional village head called *Majhi*. *Majhi* is a patriarchal hereditary social position and is an important traditional institution of indigenous governance in the *Dhimal* community. In the past, *Majhi* was entrusted with responsibilities such as maintaining social order in the village, organizing and managing the annual village *Shrejat* ritual, representing the village during the marriage processes, and mediating in quarrels or disputes among villagers, divorces, and other incidents that could potentially create conflict between villagers and villages (S. Dhimal, et al. 2010). Now many of its traditional power and social responsibilities have eroded, the institution of *Majhi* is still vibrant among the *Dhimals* and the community is making collective effort to revive the institution of *Majhi* after 1990s (Rai 2013a).

In general, the *Dhimal Majhi* families own more land in the village, and their grandparents of the present day *Dhimal Majhi* used to be the local village landlords until the early 1960s (before the Land Reform of 1964). Studies have shown that generally the village chiefs in the *adivasi* communities had transformed into landlord class after these indigenous institutions were coopted by the Gorkhali and Rana rulers for collecting expanding agricultural land and collecting land taxes for the state in the

---

wealth and power in the hands of a few by dispossessing people of their wealth and land. The *Dhimal* case I disussed here is set in a different historical context than the focus of David Harvey's (2004) analysis and characterization of the new imperialist mode of neoliberal capitalism. But the *Dhimal* case shows how possession of land, under certain political-historical conditions, actually led or could lead to dispossessions from the land. The idea of 'dispossession by accumulation' is not fully developed in this paper, but I want to highlight it in hope for further discussions on the theme. I would like to thank Stuart Kirsh and Ben Linder for pointing out this idea to me.

19th century (Caplan, 1970, Guneratne, 1996, 2002,Sugden, 2003). In the later part of the 19th century, the spur of economic activity in northern India, mainly because of the development of railway transport facilities, opened up new prospects for land colonization in the eastern Tarai (Regmi, 1978, p. 140). In order to tap these emerging economic opportunities, the state promoted more land reclamation and settlement projects (Ojha, 1982, Regmi, 1971), construction of irrigation facilities, and expansion of the revenue machinery. In 1861, the Rana rulers introduced the *jimīndāri* system as the local apparatus for state revenue administration of the *Tarai* region as well as to and to encourage "private enterprise in the colonization of large tracts of forests and other uncultivated lands whose development lay beyond the capacity of the local farmers because of the inconvenient location or paucity of capital" (Regmi, 1978, p.141). Thus, the introduction of the *jimīndāri* system which operated at the level of villages intensified the stronghold of the landlord state over the local villagers, led to increased class differentiation between villagers. *Dhimal* village heads were also incorporated as local tax functionaries or used by the *jimīndār* and *birta* holders by the early 20th century. Increasingly, the evolving state-led feudal relations and the recognition of the 'property right' of peasant cultivators in *raikar* in the late 19th century further increased the importance to state and revenue functionaries of the village head and other socially recognized community leaders or economically dominant indigenous families (Regmi, 1978).

As an alternative to becoming tenants of the state or of the landowning elites, many *Dhimal* peasants cultivated the land holdings of their village *Majhi* as sharecroppers or recipients of a fixed share of the yield. According to *Dhimals*, the village *Majhi* would hold the village land under his name and thus he would deal with tax officials and other state functionaries, while other villager members would cultivate the land and share the produce with the *Majhi*. In other words, the hardship of owning land as individual families and the state's cooption of indigenous institutions, such as the *Majhi* system, for land colonization and revenue administration in the *Tarai* impacted the existing customary social relations between *Majhi* and villagers. I argue this transformation was an effect both of the state's land tenure and revenue policies and also of *Dhimal* efforts to use their cultural institutions to blunt the effects of the oppressive state tax machinery. *Dhimal* insistence that the hardship of land tenurial relations in the past compelled them to be non-owners of land and that they found it convenient to work for or to cultivate the village *Majhi*'s lands needs an empathetic analysis.

The dominant explanation espoused by many Marxist scholars that the village heads in indigenous communities were essentially a 'landed elite class' at the village level reduces the institution of *Majhi* to an instrument of class exploitation and subjugation. But the role of *Majhi* cannot be reduced to that of landlord in any elemental sense – he was not merely a creation of the feudal mode of production dominated by Nepal's ruling elites. On the contrary, *Dhimal* claim that their *Majhi* institution predates the formation of the present-day state of Nepal. For them, the *Majhi* of the past represented an important customary institution, indispensable to the governance of *Dhimal* communal life including village ritual, marriage, and maintenance of social order. Though the *Majhi* could exercise social power to make and impose decisions on behalf of the villager, it was not a vertically ranked social position nor was it a permanent position that one could continue in without enacting and being part of locally embedded social relationships and a moral economy mediated by kinship, ritual obligation, reciprocal exchange, and other community making practices.

Many *Dhimal* peasants preferred to cultivate their village *Majhi*'s lands, not because the village head represented state power or controlled all village lands, but because the *Majhi*-villager relationships, unlike the tenancy relationship with the state or other landowning elites, were relatively egalitarian and mediated by the ethic of exchange and reciprocity embedded in *Dhimal* moral economy. These new "class" like relationships between the *Majhi* and his tenant *Dhimal* families still maintained the reciprocal relationship of production and distribution mediated through their kinship and ritual obligations. One *Dhimal* whose family in the past had cultivated the lands of the *Majhi*, the brother of his grandmother in Karikoshi village, explained to me: "During that period, we did not need

much land. Then, we did ṅot need many things and money. Our biggest
pír (Nep. worries, concern) then was the marriage of our children.
However, we used to get support from *Majhi* and others to marry off our
children. We could always rely on the *Majhi* if we needed any money and
rice. We could pay him by plowing his fields. He was our own kin." Kinship
ties, ethics of reciprocity, and the assurance that they would be helped in
times of need equally defined and structured individual families' tenurial
relations with *Majhi*. Accumulation of wealth for future investment (or
expenses) was not a salient feature of *Dhimal*'s household economy. Given
the hardship of owning land for an individual family and the social
embeddedness of their village *Majhi*, many *Dhimal* peasants found their
tenancy relationship with the Majhi more convenient and less troublesome
during that period. The views of one *Majhi*, descendant of one landlord
family from Karikoshi village succinctly captured the local challenges,
and the symbiotic and reciprocal relationships between the landlord and
the tenant families during that period. He told me in 2008:

> At that time, it was too much hardship to have land under one's name.
> There were fewer people, but land was plentiful. For example, you have
> land but no people to plow your farm. This would drown you. So what do
> you do? You need to find and *please* people to plow your farm by offering
> them money in advance, a place to live, cattle to herd, and so many other
> things. *You need to take care of these people more than your own children.*
> …….. People could not sell their crops in the market. There were no
> demands for crops in the market. People could not pay the land tax, and
> then their lands were taken by *jimīndār* and *patuwarī* and given to others.
> Not that the land tax was high. But where would people get money? There
> was a shortage of money (cash). They could not even sell their crops.
> There were few *desi* (read Indian) traders who used to come to Rangeli
> (now a town located in the southern part of Morang, bordering with
> India, former district headquarters of Morang). Not all of them could take
> their crops there to sell. Not many people knew how to make connections
> with these traders. *Dherai jhanjhāt thiyo* (It was lots of trouble).

It was obvious that *Majhi* benefitted more from retaining large tracks of land. Yet they also risked the challenges of meeting the stipulated tax requirement and pleasing the state functionaries and revenue collectors. Recall what the the landlord family told me, "For example, you have land but no people to plow your farm. *This would drown you.* So what do you do? You need to find people and *entice* them to plow your farm by offering them money in advance, a place to live, cattle to herd, and what not. You need to take care of these people more than your own children." If we were to reduce this explanation to the logic of cost-benefit analysis, given the shortage of labor and relative absence of immigrants from outside, the *Majhi,* even for a purely instrumental purpose, needed to be caring and providing. Otherwise, he had a higher chance of losing the locally available supply of labor and its loyalty, the loss of which could potentially deprive him of his land entitlements. On the one hand, this also implies that, in the absence of immigrants from outside, he provided more leverage for the tenant *Dhimals* to make their 'tenancy' relationship with the *Majhi* less exploitative. On the other hand and most importantly, in the absence of the immigrants from the outside, the local *Dhimal* peasants were able to keep village land, though it would registered be under the names of a few individuals in the state records.

I do not underestimate the issue of "tenant exploitation" in the tenancy of land under the *Majhi* (see for example Guneratne,1996). As Dhimal themselves emphasized, *Majhi*-villager relationships transcend the political economy of land tenure and economic production. In order to understand how the interface between *Dhimal* moral economy and the state's extractive political economy reconfigured the customary *Majhi*-villagers relationships, we need to consider the indigenous system of *Majhi* as, first and foremost, a culturally structured set of practices for reproducing community. Even in the present-day context, the *Majhi* system is a respected, even revered, institution in *Dhimal* society. *Dhimal* people use term '*Majhi warang*' to express their respect for and acknowledgment of their village head. The title *warang* or *warange* is used as an honorific for senior males in general and also for *Dhimal* male deities. Despite the class difference between the *Majhi* and other villagers in the past, these groups not only participated in and shared the same communal social and cultural life, they were also united by their common political subjugation and their common experience of losing ground. In the post-1950s, both families from the former landlord *Dhami* and their *Dhimal* tenants had become *sukumbāsi* (Rai, 2013a).

My discussion of the hardship and challenges of owning land, and how *Dhimals* worked to lessen that hardship shows that it was the weight of the landlord state and its oppressive tax regime that created 'landed' and 'landless' social groups among the *Dhimals*. Refusing to own land was a political choice (Scott, 2009) that many *Dhimal* cultivators made in order to avoid the hardship of being tenant subjects. But this political choice became counterproductive for *Dhimals* when property rights in land were ensured at the turn of the 20th century and when land became the most important proprietorial ownership -- a capital that one could use to access other resources. The economic expansion in the bordering region of India particularly due to development of railway transport facilities encouraged the Rana rulers to intensify the land colonization and expansion of cultivation in the eastern Tarai ((Regmi, 1978,p.140). The Nepali rulers emphasized the private reclamation of land through fiscal concession and *birta* allotments for any land colonizer, made rights to reclaimed land inheritable, encouraged the hill people to cultivate the *Tarai* land, promoted irrigation developments in the *Tarai*, encouraged colonizers to procure settlers from India and introduced the *jimīndāri* system (1861) in order to facilitate private enterprise in the colonization of large tracts of forests and other uncultivated lands in the *Tarai* during the mid-19th century (see Regmi, 1971, 1978). These renewed state interventions in land colonization brought more people to the *Tarai*, and increasingly land ownership began to shift from *Dhimals* to non-*Dhimals*, particularly to the hill "high caste" groups by the early 20th century.

After the fall of the century-long autocratic feudalistic rule of the Rana family regime (1846-1950), 'land reform' emerged as a national political project for making "modern" Nepal. Since many *Dhimals* did not have land registered under their names and lacked the necessary documentation (or access to the political power to produce such documentation), Dhimal dispossession from their land accelerated after the 1950s. When malaria was 'eradicated' in the *Tarai* during the late 1950s, the region was 'opened' for legal and extra-legal land grabbing and settlements by clearing more forests. A huge influx of immigrants, particularly from the hills, poured into the *Tarai* regions. More and more *Dhimals* lost their land through mortgage and indebtedness, sale by unfair means, development of public infrastructures like roads, schools, administrative offices, 'modernization' of agriculture (for example, the development of the tea-gardens) and other forms of 'developmental' encroachments in their territories. The Land Reform of 1964, which was implemented to impose a land-holding ceiling and to distribute the surplus land equitably to landless peasants (see Zaman, 1973) disproportionately benefitted the hill immigrants, particularly the hill "high" caste groups in the *Tarai* (see Gaige, 1975, Chaudhary, 2007 [B.S. 2064]; Guneratne, 2002). The land reform worked against the landowning *Dhimal Manjhi* landlords, who because of their lack of political connections, lost most of their landholdings. And many *Dhimals* who used to till the Majhi lands also could not secure their legal ownership of the land and became landless.

This "frontier settlement" (Shrestha, 1989) in the *Tarai* progressively dispossessed *Dhimal* from their ancestral territories and marginalized them politically, culturally and economically. Earlier studies show that the problem of landlessness had hit hard in *Dhimal* communities by the early 1970s (Dahal, 1979, Regmi, R, 1985). Frederick Gaige (1975), in his pioneering study of the marginalization of Nepal's *Tarai* people, succinctly and empathetically describes the impact of frontier settlements on the *Tarai* 'tribal' (indigenous communities) peoples. He observes:

> Until recently, the tribal people have been able to find isolation from the subcontinent's more advanced economic society in the forests of the *tarai* and other geographical peripheral regions. The surge of population into the peripheral regions and the clearing of forests to provide additional farmland have confronted the tribal people with the need to adjust to a new and essentially hostile society. Relegated as they are to the lowest rungs of the caste ladder, without the experience needed to compete for

scarce economic resources, they have generally found the adjustment process *confusing and painful. Indeed, in many cases, it has been a struggle for survival* (Gaige 1975, p. 20, emphasis added)

I argue that the *adivasi* "struggle for survival" that Gaige (1975, p. 20) so thoughtfully emphasizes should not be understood only in the sense of struggle for physical survival or only for their livelihood. It also needs to be understood as ādivāsi struggle for the continued creation and recreation of their collectivity as distinct cultural community (Turner, 1988). For *Dhimals* and other Tarai ādivāsis, their experiences of the Nepali state in the last two centuries have been about "losing ground" (McDonaugh ,1997, p. 280), which involves the collective loss of their land, culture, and their traditional hold on local political and administrative power. Hence, when *Dhimals* ask themselves and others, "We *Dhimals* are ādivāsis, but now *sukumbāsis*, why?". They advance 'land' as a political language to which Dhimal whatever their differences of class, gender, location, or generational and party affiliations feel universally connected. Stories of land powerfully concretize their shared history of belonging and dispossession in their ancestral territories. The collective experience of 'losing ground' is central to *Dhimal* political mobilization as an indigenous people. Thus, the conceptual distinction between '*adivasi*' and 'peasant' become significant to understand *Dhimal*'s encounters of the landlord state in the 19th century, their transformation into peasantry, and their experiences of 'losing ground' in Nepal's *Tarai*.

Conclusion

The majority of the *adivasi* communities in Nepal relied on farming for subsistence in greater degree or had become '*peseantized*' (Caplan, 1990) after the 19th century when their ancestral territories were annexed under the Gorkhali kingdom. We still lack historically informed studies on the transformation of these *adivasi* communities into peasant societies (Caplan, 1970, English, 1983). The changing state-*adivasi* relationships over the control of land in the *Tarai* differed from that in the hill-mountain regions. When their territories came under the Gorkhali kingdom, the

Tarai adivasi communities such as the *Dhimals* and others continuously attempted to avoid owning land under their names and evaded the tenurial relationships with the state and the landowning hill elites during the 19th and early 20th centuries.

The case of *Dhimals* that I have presented in this paper highlights that not all *adivasi* communities *relied on the 'ownership' of land on communal or individual basis even when they have 'inalienable relationships' with the nature*. The rise of the Gorkhali state in the *avatar* of the sovereign landlord in the 19th century and its colonizing project in the *Tarai* for land, labor, resource and political control radically altered the relationships between the indigenous communities, land, and the state. The 19th century land tenurial system based on the territorial sovereignty (*bhupati*) of the Gorkhali king that considered the tillers of the land -- the tenant peasants -- to be 'inferior' than the 'giver' (the king/the state) of the land. Such extractive political economy contradicted the *adivasi* moral economy practiced by the *adivasi* like the *Dhimals*. In the 19th century Nepal, to become a tenant peasant was to accept this inferior and subordinated position. Ownership of land implied the right to cultivate the land but with certain obligations to the giver of the land (the state and other land owning elites). The *Dhimali* people were essentially a non-farming people and had different understanding and relationships to the land mediated by their concept of *bhonai*. When they were compelled to reclaim the *Tarai* land for the state, their experiences of becoming the tenant subjects were only fully of hardship and subordinating. The *Dhimal* ancestors became 'landless' when they began to accumulate or own the land. Hence, many of them preferred not to own the land and continue their forest based ways of life with period farming.

In his important work, Arjun Guneratne (1996) has challenged the dominant scholarly view that the control of land is the most important guarantee of subsistence in the peasant economy. His historically grounded analysis of the changing relationships between the *Tharus* of Chitwan *Tarai* and the Nepali state over land control and tenure shows how the specific historical and material conditions such as the shortage of labor in the *Tarai*, the extractive relationship of the state with peasants, and the local manifestations of the oppressive revenue regimes in existing village social relations (landlords and peasants of the same community), that combined actually discouraged many *Tharu* peasants from owing land even when it was readily available. Instead, these *Tharu* peasants elected to work for the landlord families, and still secured their subsistence from the land through the exchange of their labor. In other words, these *Tharu* peasants opted for "voluntary landlessness" (p. 31) and became landless by choice. Guneratne's argument that "the relationship of the peasant to the land is in large part dependent on the nature of his relationship to the state" (p. 6) is particularly relevant for the *Dhimal* case as well.

As I have discussed elsewhere (Rai, 2015), the Dhimal case, also affirms Guneratne's (1996) overall conclusion. However, the Dhimal people, unlike the *Tharus*, became peasants relatively late (circa after the mid-19th century) in the history of land colonization in the *Tarai*. The *Tharus*, especially their village chiefs, had a longer history of alliance with the state rulers in Nepal (Krauskopff and Meyer, 2002) with the consequence of influencing "the form and organization of the *Tharu* society, even to its nature as a moral community" (Guneratne 1996, p. 32). Class differentiation based on land and political power had also emerged among the *Dhimals* after they became incorporated in the state land tenure system after the mid-19th century, but they lacked the kind of stratified and hierarchical social relations that Guneratne has described for *Tharu* society (see Rai, 2013a). Guneratne's analytical model is based on the concept of peasant and state-peasant relationships over land. I have approached *Dhimal* explanations of why many of their ancestors did not own land in the past by first considering the fact *that Dhimals were not non-farming people. Dhimals* had differential relationship to the land and practiced different moral economy than the landlord state. Caplan (1970, 1990) has shown how such differential relationship to the land needs to be considered for a better understanding for the transformations of *adivasi* community into peasant societies.

Along with Guneratne's analysis, I have attempted to show how James Scott (2009) analytical framework of 'escape agriculture' is relevant to understand why some communities may prefer not to own land as a political choice. Given the extractive nature of the 19th century Gorkhali state and its oppressive regimes of revenue collection, I have used the case of the *Dhimal* to argue that the practice of not owning the land also needs to be seen as a political choice that communities deployed in order to evade the state power. When *Dhimals* increasingly began to rely more on farming, they would still prefer not to own land under their names for distancing themselves with the state and its administration. Instead they would work as tenant farmers for the landed *Dhimal* families. Hence, a new form of class relations based on the landownership emerged, most likely in the early 20th century among the *Dhimals*. Because of the state's policy of promoting the traditional village chiefs as landlord and tax collectors, the *Dhimal* village head or the *Majhi* also became the local landlord. Many *Dhimals* preferred to cultivate their village *Majhi's* lands, not because the village head represented state power or controlled all village lands, but because the *Majhi*-villager relationships, unlike the tenancy relationship with the state or other landowning elites, were *relatively* egalitarian and mediated by the ethic of exchange and reciprocity embedded in Dhimal moral economy. These new class like relationships between the *Majhi* and his tenant *Dhimals* still maintained the reciprocal relationship of production and distribution mediated through their kinship and ritual obligations. But in the political and socio-economic transformations after 1950s, the *Dhimal* community, irrespective of their landholding, became a 'minority' in their own land and their experiences of 'losing ground' intensified in an unprecedented scale. My historical analysis shows that the problems of landlessness among the *Dhimals* had begun with the rise of the landlord state and its project of colonizing the land in the *Tarai* in the 19th century. Refusing to own land was a political choice that many *Dhimal* cultivators made in order to avoid the hardship of being tenant subjects. But this political choice became counterproductive for *Dhimals* when property rights in land were ensured at the turn of the 20th century and when land became the most important proprietorial ownership -- a *meeling* – a capital that one could use to access other resources. In the post 1950s, the political marginality of the *Dhimals* and other *adivasis* further dispossessed them from their land.

In this paper, I have attempted to 'bring back' the issues of '*adivasi*' and 'peasants' in the peasant studies in Nepal (Caplan, 1990, 1991, Guneratne, 1996) by focusing on the historical experiences of a non-farming indigenous community in the 19th and early 20th century. Land is central to the study of the history of changing relations between *adivasis* and the state; how this analytical focus should moves beyond the framework of peasant-class-state relationships. Similarly, the dominant framework of the 'communal ownerships' of land in the study of the *adivasis* and land is less relevant for the *Dhimals* and other *Tarai adivasis* who relied less on the cultivation of land for subsistence in the past. This paper brings new approaches to the study of the relationship of *Tarai adivasis* with the land by focusing on the interplay among the territorial sovereignty of the state, the role of the then existing ecological conditions in mediating relations among the *adivasis*, the state, and other social groups, and *Dhimals'* historical agency in resisting the extractive Hindu state.

References

Burghart, R. (1984). The formation of the concept of nation-state in Nepal. *The Journal of Asian Studies* 44(1), 101-125.

Caplan, L. (1970). *Land and social change in east Nepal: A study of Hindu-tribal Relations*. Routeledge.

Caplan, L. (1990). 'Tribes" in ethnography of Nepal: Some comments on a debate. *Contribution to Nepali Studies* 17(2): 129-145

Caplan, L. (1991). From tribe to peasant? The Limbus and the Nepalese state. *The Journal of Peasant Studies* 18.2 (1991): 305-321.

Castree, N. (2004). Differential geographies: Place, indigenous rights and 'local' resources. *Political Geography* 23(2):133-167.

Chaudharī, M. (2064 v.s [2007 A.D]. *Nepālko tarāī tathā yasakā bhumiputraharu: Nepālmā madhesa kahāṅ cha? (Nepal's Tarai and Its sons of the earth/ indigenous peoples: Where's the Madesh in Nepal?)*. Kathmandu: Śānti Caudharī.

Dahal, D. R. (2036 V.S. [1979 A.D].) *Dhimal lok jiwan adhayan (Folk life study of Dhimals)*. Kathmandu, Nepal: Royal Nepal Academy.

Dhimal, S. (2068 V.S) *Dhimāl jātiko cinārī [Introduction to Dhimal]*. Lalitapur, Morāṅg, Nepal: National Foundation for Development of Indigenous Nationalities (Nepal), Dhimāl Jāti Bikāsa Kendra (Nepal).

Dhimal, S., Dhimal G, & Dhimal, C K. (2010). *Dhimal itihas bhag-1 (Dhimal history part 1)*. Morang, Nepal: Dhimal Jati Bikas Kendra.

Diwas, T. (2035 V.S. [1978 A.D]). *Pradarsanakari Dhimala loka-samskrti (Folk cultures of Dhimal)*. Kathmandu: Nepal Rajkiya Pragya Pratisthan (Royal Nepal Academy).

English, R. (1983). *Gorkhali and Kiranti: Political economy in the eastern hills of Nepal*. PhD Thesis, New School for Social Research.

Gaige, F. (1975). *Regionalism and national unity in Nepal*. Berkeley: University of California Press.

Gray, A. (1995). The indigenous movement in Asia. In R.H. Barnes, A. Gray, $ B. Kingsbury (Eds.) *Indigenous peoples of Asia.*, (Pp. 35-58). Ann Arbor: Association for Asian Studies.

Guneratne, A. (1996). The tax-man cometh: The impact of revenue collection on subsistence strategies in Chitwan Tharu society. *Studies in Nepali History And Society* 1(1):5-35.

Guneratne, A. (2002). *Many tongues, one people: The making of Tharu identity in Nepal*. Ithaca: Cornell University Press.

Harvey, D. (2003). *The new imperialism*. Oxford ; New York: Oxford University Press.

Hodgson, B. H. (1880 [1847]). On the Kocch, Bódo and Dhimál tribes. In *Miscellaneous essays relating to Indian subjects*, 1-155. London: Trubner & Co., Ludgate Hill.

Kela, S. (2006). *Adivasi* and peasant: Reflections on Indian social history. *Journal of Peasant Studies* 33 (3): 502- 525.

Krauskopff, G. (2000). From jungle to farms: A look at *Tharu* history. In G. Krauskopff & P.D. Meyer (Eds.). *The kings of Nepal and the Tharu of the Tarai*. (Pp. 25-48) Los Angeles: Rusca Press.

Krauskopff, G. & Pamela D. M. (Eds) (2000). T*he kings of Nepal & the Tharu of the Tarai*. Los Angeles: Rusca Press.

McDonaugh, C. (1997). Losing ground, gaining ground: Land and change in a Tharu community in Dang, West Nepal. In D.N. Gellner, J. Pfaff-Czarnecka & J. Whepton (Eds.) *Nationalism and ethnicity in a Hindu kingdom: The politics of culture in contemporary Nepal* (Pp. 275-298). Amsterdam: harwood academic publication.

Ojha, D. (1983). History of the land settlement in Nepal Tarai. *Contributions to Nepalese Studies* 11(1):24.

Pradhan, K. (1991). *The Gorkha conquests : The process and consequences of the unification of Nepal, with particular reference to eastern Nepal*. Calcutta ; New York: Oxford University Press.

Rappaport, J. (2004). Between sovereignty and culture: Who is an indigenous intellectual in Colombia? *International Review of Social History 49, no. S12 (2004): 111-132.*

Rai, J. (2013a). Activism as a moral practice: Cultural politics, place makimg and indigenous activism in Nepal. PhD Thesis, University of Michigan, Ann Arbor.

Rai, J (2013b). Malaria, Tarai adivasi and the landlord state in the 19th century Nepal: A historical-ethnographic analysis. *Dhaulagiri Journal of Sociology and Anthropology 7 (2014): 87-112*

Rai, J. (2015). "Owning land was so much of dukkha in the past": Land and the state adivasi relations in the Tarai, Nepal. *Studies in Nepali History and Society 20*, no. 1 (2015): 69-98.

Regmi, M. C. (1971). *A study in Nepali economic history, 1768-1846*. New Delhi: Mañjuśrī Pub. House.

Regmi, M. C. (1978). *Thatched huts and stucco palaces: Peasants and landlords in 19th-century Nepal*. New Delhi: Vikas.

Regmi, R. K. (1985). *Cultural patterns and economic change: Anthropological study of Dhimals of Nepal*. Kathmandu, Nepal : Delhi: Sandeep Regmi ; Motilal Banarasi Dass.

RRS (Regmi Research Series) (1970). Revenue functionaries in the eastern Tarai districts. *Regmi Research Series (RSS)* 15(8):125-126.

Scott, J. (2009). *The art of not being governed : An anarchist history of upland Southeast Asia*. New Haven: Yale University Press.

Shanin, T. (1971). *Peasantsa and peasant societies: Selected readings*. Penguine books.

Shneiderman, S. (2010). Are the central Himalayas in Zomia? Some scholarly and political considerations across time and space. *Journal of Global History* 5(2):289.

Shrestha, N. (1989). Frontier settlement and landlessness among hill migrants in Nepal Tarai. *Annals of the Association of American Geographers* 79(3):370-389.

Sugden, F. (2009). Agrarian change and pre-capialist reproduction on the Nepal *Tarai*. PhD thesis, School of Geosciences, University of Edinburgh.

Turner, T. (1999). Activism, activity theory, and the new cultural politics. In S. Chaiklin, M. Hedegaard, and U.J. Jensen (Eds.) *Activity theory and social practice*. 114-35. Aarhus: Aarhus University Press.

Waddel, L. (1899). *Among the Himalayas.* New York: Amsterdam book co.

Wolf, E. (1966). *Peasants.* London: Prentice-Hall International.

Zaman, M. A. (1973). *Evaluation of land reform in Nepal, based on the work of M. A. Zaman, land reform evaluation adviser.* Kathmandu: Planning, Analysis, and Publicity Division] Ministry of Land Reforms, His Majesty's Govt. of Nepal.

Acknowledgement

I will like to thank all my *Dhimal* friends who immensely contribnuted to providing insights and information for my research on which this present paper is based. An earlier version of this paper was presented as a conference paper at the Conference on "Conceptualizing Peasentry: Understanding Peasants and Peasant Economy in Nepal" organized the by Central Department of Anthropology, TU and the Community Self Reliance Center on July 20, 2017 in Kathmandu. I like to thank the paper commentor Prof. Binod Pokharel (Central Department of Anthropology, TU) and the conference participants for their comments and suggestions. However, only I am responsbile for any shortcomings of this paper. For comments, write to janakrai2007@gmail.com

Marginalization of Tenants in Nepal: A Political-Economic Analysis

Jagat Basnet

Introduction

I would begin to discuss the situation of Nepali peasants by quoting a statement of a noted scholar, that is,"The peasants were born in fields, grew up with spades and died in the fields with almost nothing to show for them. This was generally the fate of all cultivators. These kind of problems led to underdevelopment and backwardness of each country" (Rahman, 1986 quoted byThapa, 2000, p.133). A *Newari* folk song of 1749 A.D mentions that the beauty of a tenant's wife attracted a landlord. He tried to seduce her but she refused. He compelled her to surrender by threatening to change tenants (Vaidya, 1993 quoted by Thapa, 2000, p.160). Tenants had no security in the past in terms of landholdings because landlords could evict them any time without cause and this situation remains the same even today. Media reports show that still some of the *Kamalaris* or daughters or spouses of tillers are raped by their masters or landlords (CSRC, 2011). This snapshot gives us a clear picture of the situation of land and agrarian life in Nepal since 1900 A.D. The situation has not been improved as tenant -peasants continue to be evicted on a regular basis and become more and more marginalized from the land. According to the report of the Ministry of Land Reform and

Management (MoLRM) of 2006 A.D, there were 40.4 percent tenant tillers in 1961 A.D but in 2001A.D, the percentage was reduced to 8.7 percent. Despite this dramatic drop, the government of Nepal has no data on whether they were evicted or whether land was distributed among them. According to a report from the MoLRM of March 2013 A.D, there are only 120,686 recorded tenant families and 3,323 cases have been reported to be pending at the District Land Revenue Offices or District Land Reform Offices (DLROs). According to the similar government report of 1994 A.D, there were 469,000 tenant families. But now the government report shows that only 8,995 tenants received land as per the tenancy share legal provisions till January, 2013. This has raised a question about the remaining approximately 349,000 of tenant farmers, that is, whether they were evicted by the landlords or whether they were forced to leave the landholding tilled by the tenants?. Put differently, did they sell their tenancy or were they evicted by their landlords? This chapter of the book has been written in this context which focuses on the tenants, tenancy share and land reform programme as per the Lands Act of 1964 A.D) and its subsequent amendments. The analysis presented here is soley based on the review of relevant literature supplemented by empirical obseravtions and discussions. The political-economic perspectives and their approaches have guided the analysis and discussion.

Historical Process of Marginalization

Bal Chandra Sharma's book '*Historical Outline of Nepal*' states that 60 percent of the cultivable land of the *Tarai* has been the source of personal income of some 40 to 50 ruling people. According to his explanation, around 10 percent of the land is the *Birta*, under the control of royal and *Rana* families and priests. Of the remaining, some 30 percent was distributed among big landlords and *Raikar* cultivators (Sharma, 2006). This was also proven in the report of the land investigation sub-committee, which was formed in 2007 under the Parliamentary Committee on Natural Resources, to investigate the ex-king Gyanedra's property, especially land. The report's findings state that 'over 50,926,810 *Ropanis* (2680 hectares) of land is under the name of the royal families alone'

(Parliament Committee on Natural Resource, 2007). Such ownership and control of cultivable land by the people not engaged in agriculture has a political logic, that is, "to maintain an unequal power relation in which the weak people in the chain constantly submit themselves to the strong". This pattern of ownership perpetuated a gap in caste and class division, created absent landlordism and consolidated the hold of those close to the royals as landlords. The prominent scholar, Mahesh Chandra Regmi in his book "*State as a landlord: Raikar Tenure*", argues that the state's policy, legislation and program deprive tenants and that has ultimately and adversely affected their livelihood. Most of them are from indigenous, *Dalits* and poor communities.

Among the land tenure systems described here, the most common ones are *Birta*, *Guthi* and *Jagir*. Since these land grants were mainly limited to the priests, religious teachers, soldiers, members of nobility and the royal family, they became the foundation of social and political life until political change in 1950 A.D but farming was predominantly done by tenants (Khanal,1995, p.7,Karki, 2002, p.6). Besides *Shahas*, *Ranas* and *Thakuris*, the key beneficiaries of these land grants were *Brahmins*, *Kshetries* and *Newars*. In other words, ethnic minorities and *Dalits* were not included in such land grants. A *Jimidar* could establish a settlement of peasants (even from India) in a new area on his own recommendation, and he could claim crops from the land. Selected *Jimidars* were responsible for paying a certain amount of collected land tax to the government. Since *Jimidars* had the authority to provide property ownership certificates, they used their administrative power for their personal benefits and gradually established their property ownership on wide areas of land through such practices, and became big landlords who provided a support base for the *Ranas* (Karki, 2002, p.7). The pattern of land ownership in Nepalese society favored the aristocracy. The "Bardiya Land Commission" also reported that the large-scale landlords owned 24,000 *Bighas* and small sized owners held 56,000 *Bighas* in Bardiya district in 1910. This situation was reversed 42 years later in 1952 A.D; *Jimidars* (landlords) controlled 56,000 *Bighas* while the peasants controlled only 24,000 *Bighas*. Similarly, the land occupied by the peasants decreased from 80 percent to

25 percent within the period of 15 years (1893 A.D to 1908 A.D) in the western *Tarai*. Most of the landowners were absentees which meant they were cultivating the land through the tenancy arrangements. The holdings of Prime Minister Mathbar Singh Thapa consisted of 25,346 *Bighas* land in the *Tarai* and 1797.5 *Mato Muri* (i.e 450 *ropanis*) of land in hills. In addition, he was also entitled to exact Rs. 12401 from his Khuwa village rice lands[1]. This was the common trend among the high-class families during the *Rana* regime[2] (Thapa, 2000, pp.30-31).

"There were almost 360 types of agro-commercial levies until that time and some 150 types of levies were in the effect during the time of *Ranas*. Every tenant household paid 3-5 kinds of levies regularly, for instance, the *Sauney-Fagu*, *Chardam Theki* and *Ghiukhane* levies which were paid by most of the cultivators" (Thapa, 2000, p.118). The landowning elite had imposed the cultivation terms as they wished and arbitrarily raised the rent. There was no monitoring system for this. According to Mahesh Chandra Regmi (1999), *Ranas* utilized their political power to enrich their own families by distributing the *Birta* grants. *Ranas* had the sole objective to control the political and economic power. At the end of 1950 A.D, at least one third of the total cultivated area under the *Birta* tenure and approximately three fourths of the total *Birta* land belonged to the *Rana* families (Regmi,1999, p.49). In other words, *Rajas*, *Birta*-owners and *Jagirdars* continued to occupy the role of parasitic landlords whose income from the rents was available neither from investment nor from mobilization from tax (Regmi, 1999, p.54). The ownership of land is not only the source of wealth; it is the source of prestige, power and livelihood (Beteille, 2007, p.205).

The aristocracy is, in a sense, a section of a much larger ruling class of landowners, the majority of whom constitute what might be loosely termed "gentry". Most members of the ruling class maintain close relationships with government and bureaucrats in Kathmandu often by

---

[1] 'Birta Grant to Mathbarsingh Thapa', Regmi Research Service, Year 8, No. 3 March 1976, Kathmandu: Regmi Research Institute, pp.46-49.

[2] 'Farmaisi Birta Grant to Khaga Raj Dibyawori' 1964 VS. Poka 38, SN 3797, Kathmandu: Record Section of the Department of Revenue.

means of relatives actually in government service and many have houses in the capital. Some are to be found living on or near their estates outside the Kathmandu Valley and in general even when predominantly absentee landlords, they maintain a strong local attachment, employing tenants and sharecroppers to cultivate much of the land but also frequently involving themselves in the running of the estates when they are present. Many of the large landlords are active in local and regional politics and usually carry considerable weight with the local and regional administration even if sometimes there is disagreement between the local gentry (Seddon, Blaikie & Cameron, 2002, pp. 45-46). The rent system is still traditional and feudalistic. There is no policy of central and local government. The rate of agricultural rents is decided through the bargaining of landlords. Since landlords have power, the tenants will always have to pay high rates of production. In this country, landlords are the main supporters of rulers. The rulers, therefore, work for the landlords (Thapa, 2000, pp.160-162). The tenants are forced to fulfill their duties; their rights are secondary. They were always separated from ownership and production. Land is not the property of cultivators but it was their duty to till the land. They were always asked to fulfill the needs of landlords, not the protection of tenant farmers.

Role of Land Reform Policy and Programme

Following the announcement of land reform programme in 1964 A.D as per the Lands Act, the landlords of the *Tarai* came to have audience with king Mahendra and said unanimously in a sarcastic way, "There is no possibility to feed even an elephant from the ceiling of 25 *Bighas* of land which they have been domesticating for their personal and family adventurism". Listening to the *Tarai* landlords, king Mahendra assured them that there would be no confiscation of their lands (which would exceed the ceilings set). He further assured them that there would be no implementation of land reform programme as promulgated. The main manifest objective of the land reform programme was to grant tenancy rights to the genuine tillers and regulate the agricultural rent system which was largely unrealized in the process of implementation. Unfortunately, the latent objective behind the enactment of the Lands Act was to strengthen the *Panchayat* (partyless political system with absolute monarchy introduced in 1961 A.D by the king through the dissolution of the first democratically elected government and banning of the political parties), not to implement this land reform programme. The landlords opposed the announcement of this Lands Act which largely remained in paper. Had the land reform programme been implemented, there would not have been the continuation of *Panchayat* regime. King Mahendra was so clever that he announced land reform from one side and did not urge its implementation (Rokka, 2004). At least, 1.8 million farmers should have received their tenancy rights but only 4, 69,917 tenants have received their tenancy recognition papers. Indeed, a generality of such tenant farmers have lost their rights (see Table 1 for details).

Table 1: Tenant Farmers between 1964 and 1993

Region	No. Districts	1 No. Record	4 No Record.	Total Tenants	Area of Land	3 No. Ten. Dis	Till 1993	Area Till 1993 (in *Bigha*)
Eastern	16	390277	355050	150789	243718	96230	108912	110315-8-0
Central	19	693051	544278	224086	128064	177521	217173	97485-6-4
Western	16	406678	363956	40324	21969	13530	10643	10612-11-4
Mid-Western	15	186462	179049	31715	85432	23093	28667	45569-5-4
Far-Western	9	142617	113577	23003	34031	9180	4822	11401-13-12
Total	75	1819054	1555910	469917	513214	319554	3,70,217	275384-4-4

Source: Department of Land Reform and Management, Government of Nepal, 2013.

In the process of implementing the land reform measure of 1964, tenant farmers were evicted from tilling the land and as a result, the production from agriculture was reduced which created a food deficit in the country (Rokka, 2004, p.99). This means that the landowner's rights to the land went almost unaltered (Regmi, 1976). Indeed, land reform had a more damaging effect on production and productivity because after reform, 31.2 percent of the farmers were found to be tenants. Furthermore, even after two to three decades of tenancy rights granted by the government, almost 28 percent of the households were found to be informal, unregistered tenants, especially in the *Tarai* (Zaman, 1973).

There is no consistency in data on tenancy. Different reports have provided different numbers but there are not vast differences. According to the FAO land policy analysis report, more than 1,547,000 tillers submitted applications, out of which 318,500 received tenancy certificates[3]. One of the surveys conducted by the Ministry of Land Reform in 1972 A.D estimated that 40 percent of the tenants were left out during the identification. This is inconsistent with the results of the Agricultural Sample Census taken 10 years earlier in 1962, which estimated the number of tenants at 612,000 households[4].

The *Panchayat* government introduced the third amendment of the Lands Act in 1981 A.D to achieve the original intent of the land reform to grant tenancy rights to the cultivators even if they did not have a written receipt from landlords on the eve of election of multiparty system. The amendment required the executive to appoint a committee to carry out field investigations and award the certification of tenancy if they were actually cultivating the land. This provision was only piloted in Dang district, which was never expanded to other districts. Even in Dang, this was not effectively implemented due to the power nexus of landlords (Sharama, 2012). This amendment was also repealed by the fourth amendment in 1997 A.D and as a corollary of it, unregistered tenants were subjected to injustice as there was no strong opposition by the political parties.

The land reform prorgamme was enforced throughout the country in three stages. So landlords had ample opportunity to evict their tenants whenever possible. Since tenant farmers were illiterate, they did not know the phases of land reform implementation and legal provision for tenancy. Landlords prepared legal documents to protect their lands by not granting tenancy rights to tillers. Manifestly, they wanted to continue to possess

---

[3] Ministry of Land Reforms, 1988.

[4] CBS, 1962

their excess land which was above legal ceiling set. Some of the landlords registered tenancy in the names of their own relatives, including in their own names.

According to the Lands Act, there is no possibility to bring another tenant if there is one already present (Koirala, 1987, p.3). In Surkhet, according to the District Land Revenue Officer records of 2013 A.D, it is also found that more than 10 landlords registered tenancy in their own names and in the names of their relatives or their relatives registered tenancy instead of their real tenants. Even government staff said that the Act is very much biased which provides opportunities to the landlords. The tenants preferred that the District Land Reform Office should handle their cases whereas the landlords preferred courts (Koirala, 1987, p.13).

There were various clauses of the Lands Act allowing landlords to evict tenant farmers freely. Notable among them was the mention of the "volunteer surrender clause" which stipulated that tenancy rights should not be granted to non-citizens. This gave the landlords a leeway to submit a fake document for the citizenship card of the people. According to the Lands Act of 1964 A.D, there is no provision of tenancy for foreign people. In the *Tarai*, some of the people were brought from India to protect their tenancy rights as well (Koirala, 1987). The Act mentioned, "If any real tenant leased out the land; it is not possible to grant tenancy rights to him. In this case, the tenant should surrender his tenancy rights to the landlord (Koirala, 1987, p.13). In the focus group discussions in Dang and Banke, farmers said, "It would not be beneficial to go to DLROs or courts if the tilled land is in small amount. We are tilling small plots of land. Even the landlords have given small plots to many people". The legal system of Nepal is very expensive and not easily accessible to the poor and marginalized tenant farmers. Therefore, poor people cannot afford it. In the hill districts, it takes even more than three days to visit the DLRO offices which are located somewhere else. Tenant farmers do not know the people working at DLROs and other legal people. The scribes who are from outside the community cheat them, making things even more difficult.

Regarding the transfer of tenancy rights, the Land Act states that after the death of a tenant, the land can be transferred to whoever is chosen by the landlord. In one case, the landlord started to sow the seed of conflict among brothers, forcing them to give up their tenancy. Regarding the legal loopholes, the articles of the Lands Act are not clearly defined. An ambiguous definition has given enough scope for the prejudiced judgments by the DLROs, creating enough space for the landlords to get justice from the district courts or Supreme Court. As a result, most of the landlords went to the courts but tenants could not make efforts to obtain their tenancy rights—a function of the unaffordable and inaccessible legal system. The field data have shown that most of the landlords are influential persons in their villages. For instance, when there is a dispute between the landlord and tenant, then the landlord devises how he can evict him/her from the tilling land. Most of the cases filed by landlords at the DLROs have shown the complaints on the non-submission of the grains by the tillers on time. Some of the cases have the complaints of the abuse of landlords by tenants. In isolated cases, tenants were charged as being anti-*Panchayat*. Some of the tenant farmers said that they were incarcerated from a minimum of two months to a maximum of five years on the fabricated political cases. And the release from the incarceration was possible only with the favor of local landlords who were the local politicians. Such unfavorable local social situation used to force the tillers to surrender the tilled land to the landlords who had the political, administrative and judicial nexus—a function of the landed class interests.

A tenant from Ganapur, Banke said, "His family had tenancy in the name of his grandfather. Both his grandfather and grandmother died. When he knew about tenancy rights, he made a request to his landlord. The landlord said that he had given tenancy to his grandfather, not to him. Then, he visited the DLRO. The officer said that if the landlord would be ready to consider his case, then it would be possible. Otherwise, nothing could be done. In front of big landlords, no one would speak due to his influential relationship with different people in society. Most of the landlords were affiliated with the *Panchayat* system or political parties and they had good relationships with the government officials and higher level politicians. When the *Panchayat* leaders visited district or village, they were lodged at the landlords' houses and served the varieties of foods lavishily. Therefore, such nexus would facilitate the entire process to deprive the tenants of their tenancy rights".

Traditionally, landlords have an "identity" in the community due to their land possession (because land is the source of social, economic and political power).Unlike the landlords, marginalized tenants have no "identity" in the rural areas—a function of the landlessness or limited landholdings (which is also largely marginal in the domain of productivitity).

It has been revealed from the fieldwork that most of the landlords have complained at the DLROs regarding the non-payment of agricultural rents to them on the specified time. Realistically speaking, given the fact that tenants are illiterate, they do not know the legal provisions contained in the Lands Act regarding the exact timing of payment. Participants of the focus group discussions (FDGs) in Dang and Banke districts have revealed that most of the tenants do not know the period of payment of agricultural rents. Actually, according to the Lands Act, it should be paid by March 15 (*Phalgun Masant*) of every year. Since tenants are illiterate, they cannot read the written papers handed over by the landlords. Most of the tenants shared that they received the written papers which contained mistakes in dates and names and such documents are legally unacceptable. It has also been revealed from the fieldwork that most of the landlords have left their respective villages of origin and they have appointed bailiffs to take care of the land and receive the production from the tenants. Indeed, the receipts of the payments of agricultural rents given by outside people are not legally acceptable. It has also been learnt that sometimes landlords visit the tenants in *Chaitra* or *Baishakh* (which is after March 15). The timing of such visit is not suitable for tenants because payment of the agricultural rents after March 15 has no legal validity of the temporal dimension to claim the tenancy cases.

Receiving tenancy rights has also been an uphill task. For instance, tenants who filed the cases had to wait for up to 10 years. As per the records of DLROs, none of the cases have been finalized before the duration of 18 months. It is really difficult for the tenants to wait for that long and make visits to DLROs every month. Tenants have been institutionally required to spend higher amount of financial resources in the transactional processes than the prices of tenanted pieces of land. Since landlords have power nexus, they do not need that much resource for the travel and administrative purpose. The lawyers are so expensive and are, therefore, not affordable (Koirala, 1987, p.9). At the same time, landowners claim the return of lands from their tenants and sharecroppers for self-cultivation, thus triggering widespread expulsion of tenants and sharecroppers. Thus, the end result of the 1964 Lands Act has been the failure to transform the pattern of landholding in Nepal and as a corollary of it, it has not contributed to improve the life of landless and tenant farmers of Nepal.[5] Since the imposition of ceilings and the redistribution of land were not very effective under the Lands Act, the disparity in real holdings continued to be unabated. As per the book entitled "*Nepal's Rajnitik Darpan*", Tulisi Giri said, "The government had implemented land reform programme where the landlords did not have to lose any amount of land". Indeed, such kind of land reform has not been implemented anywhere in the world. For instance, there is a proof that Yadhav Prasad Pant, a hardcore supporter of *Panchayat*, had around 1000 *bigha* of land which was not confiscated with the direction of the royal secretariat.

---

[5] At the bottom, 44 percent of the agricultural households operate only 14 percent of the total agricultural land area, while at the top 5 percent occupy 27 percent of the land. The concentration index for agricultural land is 0.54 reflecting highly uneven distribution of farm land (Shiva Sharma, "Land Tenure and Poverty In Nepal"). Paper presented in WDR-2000 consultation meeting organized by the World Bank, April 4-6, 1999, Dhaka.

Unjust Land Ceiling and Weak Implementation

The government specified the land ceiling through the land reform measure of 1964, which was glaringly skewed between the landlord and tenant, that is, 2.7 hectares for tenants and 18.4 hectares for landlords. This was indeed an unjust ceiling system. Similarly, the land ceiling of high hill and *Tarai* was also unjust for the high hill people (because there farmers can grow one crop whereas *Tarai* people can harvest three times/ year). The Table 2 shows the details of land ceiling for landlords and tenant families in the different regions.

Table 2: Land Ceiling

Area	Ceiling in Agricultural Land (in ha)	Ceiling in Homestead Land (in ha)	Total Land Ceiling for Landlords (in ha)	Total Land Ceiling for Tenants (in ha)
Tarai and Inner *Tarai*	16.4	2.0	18.4	2.7
Kathmandu Valley	2.7	0.4	3.7	0.5
Hill region and other	4.1	0.8	4.1	1.0

Source: Lands Act 1964 A.D, Government of Nepal.

These restrictive provisions are not applicable in the cases of government land in possession of medical, educational and religious organizations and land used for industrial and specific agricultural purposes and for cooperative farming (Zaman, 1973, p.12). Since there was power nexus, most of the landlords converted their lands for agricultural cooperatives or industrial farms, which were in excess of the land ceiling.

Tenant farmers who received the land under the government's land reform programme could not pay the fixed land price in one installment. Actually, land was not distributed but rather it was sold. The price of land ranged between Rs. 140 and Rs. 280 per *Bigha* as per the quality of land. Those who bought the land with the government loans had to pay in 10 installments and the interest rate was 10 percent. But the interest rate was calculated on a daily basis and the tenants could not pay the interest rate on time which forced them to take loans with usurious interests. Their financial inability to repay the loans, in turn, resulted in handing over the lands to the landlords. There was no possibility of buying the land by the landless farmers who were already buried into "debt trap" of the landlords. Thus, in the name of tenants and landless farmers, land was bought again by the landlords and government officials. The land reform of 1964 A.D contributed to the emergence of new landlords. According to the Badal Commission constituted in 1995 A.D, only less than two percent of the total land was acquired through the process of land reform and 1.5 percent was distributed because the ceiling provision was not introduced in all districts at once. During the time interval of the three-phase implementation of the land reform, landlords were able to adjust their holdings while maintaining the ceiling through sale or fake transfers among close friends or relatives[6]. The government policy on land ceilings provided loopholes to distribute land among their family members as well as other individuals. The Act allowed landowners to distribute land, exceeding the ceiling, to their children who have attained the age of 16 resulting in the government failure to confiscate the land from the landlords. As indicated above, the Lands Act was implemented in three phases which gave ample time to the landlords for the anticipatory transfers of lands in the names of members of their families and their faithfuls which was exclusively done to adjust their excess holdings as per the limit of government land ceilings. The donation to public educational institute by the landlords having excessive land was permitted even after the commencement of the ceiling clause[7] (Sharma, 2012).

Exploitative Rent and Eviction

Although land reform programme was ostensibly designed for poor people, the Lands Act of 1964 promoted exploitative tenancy relationships. Indeed, this authorized the landowners to charge rents or collect shares of production thereof from the tenants of up to 50 percent of the main

---

[6] High Level Land Reform Commission (Badal Commission) Report, 2051, p. 86, see also High Level Scientific Land Reform Commission, 2067, p.13.

[7] Sec. 7 (5)

annual crop yield (High Level Land Reform Commision, 1995, Sharma, 2012). The 50 percent crop sharing due to having ownership on land is not a fair provision. Compared to Nepal, the land reform initiatives in different states of India have been better which have reduced the crop sharing to maximum one third of the annual crops (Sharma, 2012). Even with natural calamities, there is no production and therefore, tenants cannot submit the agreed production. If tenants cannot pay the crops as mentioned in the bond paper, they would be arrested and sent to jail and thus they would be compelled to give up their tenancy rights. Many reports have indicated that the Lands Act of 1964 generally has reinforced the interests and position of the landowning classes as the rents were fixed at high rates and landowners had no obligation for production or tenants' welfare (Zaman, 1973, Regmi, 1976). The authority of rent fixing is given to the landlords, which is connected with the eviction of tenants. The Act gave ample spaces to landowners to file petitions to expel tenants. Although tenants have paid the rents, they could not ask for the payment receipts—a function of their illiteracy and their subaltern position. As a result, landlords have enough spaces to file cases on the fabricated case of the non-payment (due to the non-issuance of payment receipts by the landlords). The land reform office, in such condition, may issue an order to expel such tenants (The Government of Nepal, 1964).

The Act has also given two other conditions on which a tenant can be expelled. Firstly, if a tenant has knowingly done any act or has failed to take reasonable care of land, consequently, the value or crop would decrease. Secondly, if the tenant has not cultivated for a circumstance beyond the tenant's control or has neglected in the cultivation for a year (The Government of Nepal, 1964). Out of three grounds of expulsion, the first one is too easy for a landowner to prove, as s/he is the one who denies issuing a receipt. The provisions on reasonable care and the negligence in the cultivation may also be unreasonable accusations to expel the tenants. There were 80,719 complaints lodged in land reform offices by the landowners to evict the tenants only on the accusation of rent-related obligation provided by the Act (High Level Land Reform Commision, 1995). As landlords lodged the applications for eviction, most of the tenant-farmers lost their tenancy rights, livelihood security and this has also contributed to involving them in the Maoist conflict (Sharma, 2012). For example, an octagenarian Budhai Kadriya, a tenant from Ganapur, Banke said, "Each year, the landlord files cases at court to evict tenants on the pretext of the non-payment of rents. I had also filed the case eight years ago against the landlord, and I am still visiting the DLRO each month to get the justice. My son used to work as bonded labourer in the landlord's home but later on, he changed his job and the landlord wanted to evict me from tenancy rights. He also increased the rent amount which is more than 60 percent of the total production. I produce paddy and wheat which I have to submit to the landlord. I am taking loans for my subsistence. One of my son works at Nepalgunj market at a very low salary due to low level of level education. He provided me with Rs 1000 which is being used by me to visit the DLRO and buy very simple food. Other illiterate tenants are facing the same kind of problems". Expectedly, the offices of land reform and Village Development Committees (VDCs) could be the offices working for tenant farmers but they work for the landlords. In other words, tenants and peasants of Nepal share their production not only with the government, the landowning aristocrats and bureaucrats, but also with the functionaries employed to collect the share (Regmi, 2002, p.177). If people buy something, they ask for the receipts. But if a tenant asks for the receipt of the payment of agricultural rent, then he is evicted. Generally, the agreement for tenany and sharecropping is concluded between trusted partners.

In the Name of Land Reform

According to M.A. Zaman's report to the government of Nepal and as shown in Table 3 of the report, 4.1 percent of households occupied 27.5 percent of agricultural landholding (if we consider houseolds with 10 ha+). Conversely, the *Panchayat Smarika (a Mouth Piece Journal of Panchayat)* showed that one of the major progresses of *Panchayat* period was land distribution to the tillers and landless people (Dungel, n.d). Similarly, 63.5 percent of farmers held only 10.5 percent of land, which was less than 1 hectare. This clearly exemplifies the skewed land distribution in Nepal and how genuine peasants are being marginalized from tilling the land. According to the Human Development Report (UNDP, 2004), 5 percent rich households possess 37 percent agricultural land, which shows the same kind of situation and skewed distribution of land.

Table 3: Size of Landholding in 1972

Size of Holdings	Households in %	Area in %
Less than 1 ha.	63.5	10.5
1-3ha	19.5	18.0
3-5 ha	17.1	12.0
5-10 ha	5.8	21.0
10-15 ha	2.1	11.0
15-20 ha	0.9	7.0
20-30 ha	0.5	5.5
30 and above	0.6	15.0

Source: Zaman, M. A. (1973). *Evaluation of Land Reform in Nepal,* Kathmandu: Ministry of Land Reform and Management, Kathmandu, Nepal, Table -I, p.6.

Although *Panchayat* showed major progress regarding land reform, the big landlords had 19 hectares on average in 1961 but in 1981, it reached up to 22 hectares after the land reform. This reveals that there was an increase in landholding among a limited group of people instead of the empowerment of tenant farmers in the regime of landholding/landownership (Rokka, 2004, p.85).My fieldwork in Banke, Bardiya, Dang and Surkhet districts has revealed that the skewed land distribution and marginalization of tenants from the land were the fucntions of state's policies and programmes and its apparatus from district to the national level. The land reform programme was introduced to protect the *Panchayat* system rather than to distribute land to the tillers. The analysis of above table clearly shows that the skewed distribution and marginalization from the land in Nepal has created absentee landlords/absentee owners and tillers.Obviously, there is the emergence of two classes, namely, landowning and cultivating. The cultivating class has been working for the landowning class people and their livelihood is dependent upon them. The landowning class has been also controlling the labour and tenant farmers. The statement presented below shows the situation of landlordism in *Panchayat* period:

"Harihar Gautam, from Rajapur, Bardiya had controlled over 18,000 *Bigha* of land (12,000 hectares). In 1963, King Mahendra met him in Bardiya and asked him: "What did he want?". He said, "I have enough land and therefore, I do not need anything". When King Mahendra announced the land reform programme in 1964, he had a heart attack. He was taken to Lucknow, India for his treatment in a privately hired plane but unfortunately, he died on the way. He also had large plots of land in Banke and Kapilbastu districts". Madhav Dhungel, Nepal Weekly, Sunday, 30th Jestha 2061 (12 June 2004), Kantipur Publication, Kathmandu Nepal.

In Rajapur of Bardiya district, there were 22,000 hectares of fertile land controlled by a few landlords who used to live in Nepalgunj and Kathmandu. They would come to Rajapur only during the winter season

Table 4: Number and Area of Landholding
by Tenure and Size in 1981/2

Size of Land-holding (in ha)	Number of HHs	% of HHs	Area covered in ha	% of area cover	Average Households' Land-holding (in ha)
No land	8224	0.4	-	-	-
0.0-0.5	1099677	50.1	161999	6.6	0.15
0.5-1.0	355420	16.2	264930	10.8	0.74
1.0-2.0	379051	17.3	490413	19.9	1.29
2.0-3.0	156961	7.2	379590	15.4	2.41
3.0-4.0	77228	3.5	266513	10.8	3.45
4.0-5.0	42441	1.9	189159	7.7	4.45
5.0-10.0	60082	2.7	388679	15.8	6.46
Above 10	14872	0.7	322434	13.0	21.68
Total	2193956	100	2463717	100	-

Source: CBS, National Census of Agriculture, 1981/2 (Kathmandu, CBS 1985a, p. 2)

Ghimire, K.(1998), Forest or Farm? The Politics of Poverty and Land Hunger in Nepal, Table 1, p. 15.

Note: HHs=households

to protect themselves from the cold. They visited Rajapur, Bardiya to hunt wild animals, rape *Tharu* and *Badi* girls and be engaged in other adventures. But nobody dared to speak against their behavour and illegal works. If anybody spoke against them, their family members or relatives of that particular person would be arrested by the police and detained in the custody. According to the media reports, the Maoists seized large plots of land in Rajapur of Bardiya district between 1996 A.D and 2006 A.D which is indicative of the existence of a large number of absentee landlords.

According to above Table 4, 53 percent of landlords with more than 4 ha had 36.5 percent of agricultural land in 1981 when King Mahendra's land reform programme had already been implemented. During the field discussion in Bardiya, farmers said that the political parties did not mobilize the citizens to implement the land reform programme; instead they asked them not to get involved in the King's land reform programme. Moreover, they were asked to organize the parties' 'Farmers' Associations'. These parties also assured these citizens to initiate/institute land reform after the establishment of a multiparty system in the country. However, this was never implemented even after democracy was established in the year 2046 B.S (1990 A.D); rather they changed their position on land reform and strengthened their coordination and partnership with landlords. According to the Annual Progress Report of the Parliament Committee on Natural Resources and Means, 2007 A.D, King Birendra registered 465-4-1 *Bigha* of land in Latikoili, Surkhet in 1978 A.D (B.S 1 Aashad, 2034). The *Tharus* did not receive any land although they were farmers, tenants and sharecroppers who made the land cultivable. This example shows that land was measured and registered in the name of royalists, zonal administrators, district level chiefs and members of National *Panchayat* but it was not provided to the tillers. Further, the empirical data and the existing studies show that cultivators are marginalized from the land which, in turn, has further weakened their livelihood and marginalized them in their education.

During the 30 years of Panchayat regime (1961-1990 A.D), the Palace Secretariat had became the nerve centre for the administrative and the political structure of Nepal, even though it's dominant policy and decision-making was not defined via law or the constitution of this country. The function of the Palace Secretariat could be closely compared to the previous, all-powerful function of the hereditary *Rana* Prime Minister's Office. It not only worked as Secretariat of the Palace but it was functioning as a secretariat for the royal government. The royal family and upper classes were always in decision-making positions, leaving the poor at a disadvantage as it tended to occur in this kind of power nexus (Seddon, Blaikie, Cameron, 2002, p.47). Official records have also shown that the royal family is tenant of around 445 *ropanis* land in Kathmandu valley. According to the definition of tenants, farmers tilling the land of the landlords are known as the "tenants" (Progress Report of Parliament Committee on Natural Resource and Means, 2007).

In 1996 A.D, there was a decision on elimination of dual ownership. The Lands Act Fourth Amendment of 1997 A.D favoured only the registered tenants and there was injustice for approximately 500,000 unregistered tenant farmers (Badal High Level Land Reform Commission, 1995). But no one filed a case contesting this and there was no mass mobilization against this. However, on the contrary, in the year 2001 A.D when the land ceiling was lowered, cases were filed at the Supreme Court mentioning the violation of property rights and human rights. There was no violation, when around 0.5 million-tenant farmers lost their land rights. No one talked about the property of these tenant farmers but there was a massive discussion regarding the property rights of the rich and landlords who actually had made the land in their names without any investment via the backdoor due to their power nexus. Those who are large and absentee landlords have appointed local agents at their respective villages to collect agricultural rents to protect their land. These local agents always pass information to the landlords. Some of the landlords also appointed their militia to take action against the tenant farmers or sharecroppers as needed.

The aristocracy is, in a sense, a section of a much larger ruling class of landowners, the majority of whom constitute what might be loosely termed "gentry" (Beteille, 2007). Most of the members of the ruling class maintain close relations with the government and bureaucrats in Kathmandu, often due to having relatives in government service. Many have houses in the capital or district headquarters or major cities of Nepal; however, some are found to be living in or nearby the Kathmandu Valley. But in general even though they are absentee landlords, they maintain a strong local attachment, employing tenants and sharecroppers to cultivate much of the land. They also frequently involve themselves in the running of the estate when they are present. Many of the large landlords are active in local and regional politics and usually carry considerable weight with the local and regional administration even if sometimes there are disagreements between the local gentry (Seddon, Blaikie, Cameron 2002, pp.45- 46).

Table 5: Distribution of Land under Land Reform in Hectares
(in 1964 A.D)

Region	Cultivated Area	Above Ceiling	Confiscated	Area Sold
Tarai-Eastern	167247	9153	4746	3380
Tarai –Central	364879	6645	943	377
Tarai-Western	133382	25173	23880	18723
Inner *Tarai*	658560	6221	1053	559
Sub-total	1518785	58913	33676	23524
Kathmandu Valley	42,577	7062	149	54
Other Hill Districts	764,638	405	6	-
Total	2,326,000	66380	33825	23588

Source: KC, Ram Bahadur (1986). *Land Reform Progress and Prospect in Nepal.* Research Report Series No.2 July 1986, HMG-USAID-GTZ-WINROCK Capacity Building Project, Table 3, p. 7

It was estimated that 600,000 hectares of land would be acquired from land reform to make redistribution of land to landless and tenant farmers. The combined area of all land holdings exceeding ceiling levels and available for redistribution was only 3 percent, of which less than 1.5 percent was legally appropriated and only 1 percent was distributed (KC, 1986, p.5). This was not a land reform programme; it was only sharing of the land among the rich families.

According to the informants from Dang, Banke, Bardiya and Surkhet districts, the state provisioned two documents, namely, citizenship cards and recommendation from the Chairperson of *Panchayat* or VDC but these were fully controlled by the landlords. These two documents were the preconditions for claiming and receiving tenancy rights. These two documents were only possible for them after the declaration of Nepal as a republic state but by then, all of the lands were totally controlled by the landlords or rich people. The literature has also shown that in the Kathmandu Valley, the average size of the landlord's household is 0.41 hectares whereas it is 33.33 hectares in western *Tarai*. Between are two extremes of land holdings, eastern *Tarai* 18.57 hectares and western hills 8.29 and eastern hills 7.34 hectares (Zaman, 1973, p.28). These data show that although the land reform programme was announced by the government, it was weakly implemented. Neither the state had developed any strong monitoring mechanism to track the changes of the process of the implementation of 1964 Lands Act.

Table 6: Landholding Size (in 1973 A.D)

Size of Holding Class	Landlords		Owner-cultivators		Owner-cum-tenants		Tenants		Total	
	H	A	H	A	H	A	H	T	H	A
Less than 1 hectares	15.0	0.3	66.6	12.5	70.0	21.5	50.0	9.0	63.1	10.6
1-3 hectares	6.4	0.6	18.5	20.0	17.4	26.2	32.6	40.0	19.4	17.9
3-5 hectares	16.0	3.7	6.0	13.2	7.6	20.4	10.5	22.1	7.1	11.0
5-10 hectares	7.4	2.8	6.4	35.3	3.5	15.6	5.5	20.3	5.8	21.4
10-15 hectares	6.4	4.5	2.1	15.0	1.5	12.3	1.4	8.6	2.5	11.4
15 -20 hectares	16.0	16.0	0.4	3.5	-	4.0	-	-	0.9	7.0
20-30 hectares	13.8	18.0	-	0.5	-	-	-	-	0.5	5.5
30 & above	19.4	54.1	-	-	-	-	-	-	0.7	15.2
Total	100	100	100	100	100	100	100	100	100	100

Source: Zaman, M. A. (1973). *Evaluation of Land Reform in Nepal*. Kathmandu: Ministry of Land Reform and Management, Kathmandu, Nepal, p. 29, computed from table 10 of Annex III.

Note: Percentage is calculated on the basis of the sample survey result. H=Households, and A= Area.

The above Table 6 has shown that 0.7 percent households controlled 15.2 percent of agricultural land whereas 63.1 percent of the households had possessed only 10.6 percent agricultural land. This shows that there was no land reform on behalf of tenants. The rent system was still traditional and feudalistic. There was no policy of central government and local government on the rent system. Rent was decided on the bargaining capacity of the landlords. This is also true at the moment. Indeed, it is also a matter of power. But, as the tenants and sharecroppers are powerless and economically poor, they do not have bargaining power. Since the landlord has power, the tenant always pays a high rate of agricultural rent (Thapa, 2000, p.160). According to Zaman, the government never compiled the full set of cadastral records necessary for the programme and it stood idle while most landowners circumvented the laws by subdividing holdings among relatives or friends. Complicated land legislation cannot be followed and understood by the indigenous people and the complicated legal process is not friendly to poor people to ensure their rights over land.

Table 7 shows that the government confiscated only 47805-4-0 *Bighas* of land and distributed only 43685-15-15 *Bighas*, which was expected to be around 1 million *Bighas* of land. Some of the land was transferred to fake institutions controlled by the landlords. The 1964 land reform measure neither evolved scientifically nor was implemented sincerely. At a time when land reform was implemented through the Lands Act of 1964, 65 percent of poor peasants had 15 percent of land as opposed to 39.7 percent of land possessed by 3.7 percent rich peasants and feudals (CBS, 1961, Bhattarai, 2003, p.125). After the land reform, the number of affected landlords was only 9136 with 50,580 hectares of land recorded as above ceilings.Out of this, 32,331 hectares was acquired, of which only 64 percent was redistributed (Zaman, 1973). As a result, out of the total cultivated area, 9.9 percent rich peasants and landlords owned 60.8 percent of the land after land reform (CBS, 1971, Bhattarai, 2003, p.126-127). According to the media report during the period of 1964-67, 1992-1995, and 2006-2008, many landlord families divorced or separated on papers to save their land. Many families distributed their land to their relatives and shared the property with their family members.

Table 7: Government Land Reform and Implementation of
Land Ceiling in Hectares (between 1964 and 1967 A.D)

Regions	Districts	Received Over Land Ceiling	Confiscated But Not Shown	Total confisc- ated Land	Distribution of Land	No. of Land Receiving Families			
						Tenants	Landless	Institutions	Total
Eastern	16	5694-1-14	627-0-1	6322-1-15	5480-13-16	312	112	0	424
Central Region	19	2966-2-1	219-0-0	3185-2-1	1838-6-19	1966	126	622	2714
Western region	16	3414-11-8	104-0-0	3518-11-8	2544-13-14	1743	322	7	2075
Mid-West	15	25542-9-0	5973-0-0	31515-9-0	30588-18-16	10557	4874	10	15441
Far- West	9	2538-19-16	725-0-0	3263-2-10	3233-2-10	501	665	0	1166
Total	75	40,157-3-19	7,648-0-1	47,805-4-0	43,685-15-15	15,082	6099	639	21,820

Source: Department of Land Reform and Management, Government of Nepal, 1993, Register Record, p. 18.

The role of land revenue, land survey and land reform offices remained only for the fragmentation of land or the transfer of the land from one place to another. They also distributed the land among the elites. According to the rule, land in the rural areas should come under new cadastral survey in 20 years' time. But most of the surveys are carried out in urban areas because the officials can earn more money. The value of land in urban areas leads to bribes. For this study, I had a discussion with former Secretary of MoLRM. He said, "Surveyors are the problems who create conflicts among people. The government staff makes all mistakes but people should go to the courts. There is no punishment to government staff. Poor people cannot spend money, and lose their land. The MoLRM is the most corrupt office and its staffs are hell-bent on earning money through the foul means. There is no strong monitoring system". Mr. Jaganath Acharya wanted to implement the land reform genuinely but he was sacked from the Minister portfolio. Eventually, he was even compelled to leave the Nepali Congress due to the persistence of his commitment (Basnet, 2010). In 2001 A.D, Deuba government lowered the land ceiling but it was not implemented at all. This gave the new definition of "family" which allowed the persons over 16 years even within a single family to own the land. Sailaja Acharaya, ex-Deputy Prime Minister in the Nepali Congress government, said, "This announcement of new land ceiling would incite the landlords of *Tarai* to revolt against Nepali Congress if it is implemented". The aftermath of this announcement of lower ceiling triggered the increase of large number of fake petitions on the family separation for materializing the property separation.

Recapitulation

The above analysis has clearly demonstrated that the goal of appropriating the excess land above ceilings and redistributing it to the landless peasants and tenants was largely under-achieved. The three-phase implementation strategy of the government indicated that the entire purpose of the much-trumpted land reform measure was diatmetrically phony because such strategy allowed the landlords to devise their tricks for the anticipatory transfers of excess land above the ceiling in the names of their family members, near and distant relatives, and their faithfuls. The Lands Act also allowed to transfer the land in the name of industries and educational and religious organisations instead of transferring it in the names of the tenants. There was higher concentrartion of landholdings among few landlords in western *Tarai* but land reform programme started from eastern part of Nepal where there was smaller concentration of the landholdings than in the western *Tarai*. Indeed, land was not redistributed but rather sold (but buying land was not within the financial capacity of tenants). The local *Jimidars, Talukdars* and *Chaudharis* (the traditional land revenue collection functionaries of the government) and their educated offspring had turned to be the government staff who never had any intention to implement the land reform programme. The excess land of the hardcore *Panchayat* cadres and leaders was not confiscated. Indeed, the provision of tenancy rights under the 1964 Lands Act was only nominally implemented.

References

Basnet, J. (2008). *Land for growth and poverty alleviation*. Unpublished Paper

Beteille, A. (2007). *Marxism and class analysis*. Delhi: Oxford University Press, India.

Bhattarai, B. R. (2003). *The nature of underdevelopment and regional structure of Nepal: A Marxist analysis*. Delhi: Adroit Publishers, India.

Camron, J, Seddon, J.D, and Blaikie, P.M, (2005). *Nepal in crisis: Growth and stagnation at the periphery*. Delhi: Adroit Publishers, India.

CSRC (2011). *Land tenure and agrarian reform in Nepal*. Kathmandu: CSRC.

Department of Land Reform and Management (2013): *Record of land ceiling and tenant farmers from 1964-1967*. Kathmandu, Nepal: Ministry of Land Reform and Management.

Dungel, M. (n.d). *Land and land reform in Nepal*. Kathmandu, Nepal: CSRC.

Ghimire, K. (1998). *Forest or farm? The politics of poverty and land hunger in Nepal*. Delhi: Oxford University Press.

Government of Nepal (2007). *Annual progress report of parliament committee on natural resources and means, 2007*. Kathamndu Nepal: Government of Nepal.

Government of Nepal (1964). *Lands act*. Kathamndu Nepal: Government of Nepal

High Level Land Reform Commission (1995). *A report submitted to the HMG/ Nepal by high level land reform commission.* Kathamndu: *Paraibi Publication,* Nepal.

Karki, A. (2002). Movement from below: Land rights movement in Nepal. *Inter-Asia Cultural Studies* (Volume 3, Number 2, August 2002).

KC, R. B. (1986), *Land reform progress and prospect in Nepal.* Research Report Series No.2 July 1986. Kathamndu: HMG-USAID-GTZ-WINROC Capacity Building Project.

Khanal, D. N.(1995). *Land tenure system and agrarian structure of Nepal.* Rome: FAO.

Koirala, B.P. (1987). *Economics of land reform in Nepal: Case study of Dhanusha.* HMG-USAID-GTZ-IDRC-FORD_WINROCK Project Strengthening Institutional Capacity in the Food and Agriculture Sector in Nepal.

Ministry of Land Reform (1998). *Records of tenant farmers.* Kathmandu: MoLR.

Ministry of Land Reform (2006). *Records of tenant farmers.* Kathmandu: MoLR.

Rahman, A. (1986). *Peasant and classes: A study on differentiation in Bangladesh*: Delhi: Bikash Publishing House.Regmi M. C. (2002). *Nepal: An historical miscellany.* Delhi: Adroit Publishers.

Regmi, M. C. (1978). *Land tenure and taxation in Nepal.* Kathmandu: Ratna Pustak Bhandar.

Regmi, M.C. (1976). *Land ownership in Nepal*: Delhi:Adroit Publication.

Report on Parliamentary Committee on King's Land (2007). *Progress report of parliament committee on natural resources.* Kathamndu: Government of Nepal.

Rokka, H. (2004). Nepali daridrata ra sanrachanagat samayojan karyakram. B. Gautam, J. Adhikari and P. Basnet (Eds.) *Nepalma Garibiko Bahas.* Kathmandu: Martin Chautari, Nepal

Seddon, D., Blaikie, P. and Cameron, J.(2002). *Peasants and workers in Nepal.* Delhi: Adroit Publishers, India.

Sharma, P. (2012). *Review of land-related legislation in Nepal in preparation of an integrated land law.* Kathmandu: FAO, Kathmandu, Nepal.

Sharma, R. (2006). *Nepali shram: Sangkatma shram Shakti* B. Gautam, J. Adhikari and P. Basnet (Eds.) *Nepalma Garibiko Bahas.* Kathmandu: Martin Chautari, Nepal.

Thapa, S. (2000). *Historical study of agrarian relations in Nepal (1846-1951).* Delhi:Adroit Publishers, India.

UNDP (2004). *Human development report 2004.* Kathmandu Nepal:UNDP.

Upreti, B.R., Sharma, S.R., and Basnet, J. (2008). *Land politics and conflict in Nepal: Realities and potentials for agrarian transformation,* Kathmandu: Community Self Reliance Centre (CSRC), Swiss National Centre of Competence in Research (NCCR) North-South and Kathmandu University.

Vaidya, T.R. (1993). *Prithivi Narayan Shah: The founder of modern Nepal.* New Delhi: Anmol Publication.

Zaman, M. A. (1973). *Evaluation of land reform in Nepal.* Kathmandu: Ministry of Land Reform and Management, Nepal.

Peasantry and State-People Relations in Food Security Governance: Exploring the Linkages from Gender Perspective

YAMUNA GHALE

Defining 'Peasants'

In general, a 'peasant' is defined as "a member of a traditional class of farmers, either laborers or owners of small farms, especially in the middle ages under feudalism, or more generally, in any pre-industrial society" (https://www.google.ch). In Europe, peasants were divided into three classes according to their personal status: slave, serf, and free tenant. In Nepal, peasants are not defined as such but understood as a synonym to resource-poor small farmers and farm-based laborers. According to Nepal Living Standard Survey 2011, Nepal still occupies 76 percent of its population in agriculture. Out of 26 percent of the population below poverty line, majority are those resource-poor farmers and or farm-based laborers. Therefore, it is crucial to understand the nexus between poverty, peasantry and empowerment for their meaningful participation, representation and influence in the policy processes and its operationalization. It should serve the purpose to make the state accountable to its citizens, especially to peasants.

Poverty, Peasantry and Exclusion

Poverty is a complex phenomenon. Policy makers often define 'poverty' guided by the economic indicator irrespective of its social, cultural and political aspects. The intersectionality of those crosscutting concerns and its application are largely being guided by the person's attitude and behavior in translating it into the policies, norms and practices. Because of such practices, citizens' engagement in specific peasant's role in policy processes has largely been ignored. In Nepal, poverty reduction has been the core objective of all periodic development plans since the 1950's. Poverty, thereafter, has reportedly decreased to 26 percent by 2011. Among the population suffering from poverty, highest incidence is among the farming population (27.2%) and those who have no land or 1-2 hac of land (CBS, 2011). The latest poverty status of Nepal has been reported to be 21 percent (NPC, 2017). Likewise, malnutrition is most prevalent among the population from mountain and hills of far and mid-western region with 48 percent and 75 percent, respectively (FAO, 2010). It shows that understanding the intersectionality of poverty and state's response to different needs and entitlements associated with gender, class, geography and caste/ethnicity are still largely overlooked. Therefore, it is important to understand poverty and responding policy frameworks in relation to an individual's context as rights holders and state's responsibility as the duty bearer. It will thus help to achieve mutual objective of citizen's empowerment and state's accountability. However, structural issues that govern poverty and sustained peasantry are to be responded. In this context, establishment of systematic and inclusive governance has remained as a far-reaching objective so far. The context thus requires policy frameworks to facilitate the transformative processes in removing all structural barriers including psychological well-being of citizens and their potential contribution to the overall socio-political and economic growth of the country.

Poverty as such is a manifestation of unequal power relation. History has been a clear evidence of the fact that power relation is always determined by gender, class, caste and geography-biased disparities. The rooted disparities are also reflected in the overall development priorities and policy frameworks. In this process, peasants who are the custodians of natural resource management, indigenous knowledge and skills and local food system supporters get systematically excluded from the formal policy processes. This exclusion process undermines psychological wellbeing of peasants who are then naturally disempowered. In the process of empowerment, self-deception also plays crucial role. Self-deception is defined as *the action or practice of allowing oneself to believe that a false or unvalidated feeling, idea or situation is true (https://www. google.com.np/search?source=hp&q=what+is+selfdeception&oq=what+is +self-decepe&gs_l=psy-ab)*. Hippel and Trivers (2011) argue that a person can be both deceiver and deceived. Therefore, it is important to understand the situation that facilitates the process in such a way that self-deception can lead to self-enhancement. This process will allow a person to display more confidence, which has great social advantages. Otherwise, self-deception manifests in many different forms of psychological processes. Therefore, in achieving the empowerment, a peasant in general and a woman peasant in particular need proper attention for their voices, choices in a safe way.

By these multiple reasons, peasants with economic deprivation and marginalization have been systematically excluded from the process of participation, representation and influence in decision-making processes. While peasants are excluded, participation of women among peasants is still a far-fetched dream. Such exclusion process does not only restrict them to be empowered but also hinders the process of broader political, economic and social transformation. It proves that poverty is a state of non-acceptance, exclusion and discrimination. Moreover, women peasants face multiple forms of discrimination, inequalities and violence, which, in turn, results in disempowerment. Figure 1 below depicts the circle of multiple exclusions for women. The report of Ministry of Agricultural Development (MoAD), FY 2071/72 (2015/16) shows that overall agricultural development programme supports at district level served 8.7 percent of *Dalits*, 36.4 percent of *Janajatis* and 54.9 percent others. Gender-wise, a slightly more than half of the major programmes (such as general agricultural extension services and horticultural programmes) have reached women for general agricultural extension services and horticultural programmes. Income poverty-wise, the beneficiaries of Ministry's programme are 24.9 percent disadvantaged groups and 75.1 percent are others. The report, however, does not provide any information on how the participation in those programmes was contributing to the overall transformation of structural barriers such as access to and control over productive resources, peasant's participation in decision-making roles of different executive bodies and how they are encouraged to participate in overall state planning process to assert their entitlements.

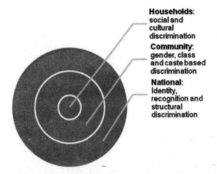

Households: social and cultural discrimination

Community: gender, class and caste based discrimination

National: Identity, recognition and structural discrimination

Figure 1. Multiple exclusions to women

Therefore, poverty is a function of systematic exclusion and deprivation from opportunities, spaces, voices and choices to influence the public policy processes safely. Therefore, it is important to understand the nexus between poverty, power relation and governance system as a whole. It will help in the transformation of discriminatory policies and practices that are required through dual approaches of empowerment of peasants and state being accountable.

The situation seems prevalent throughout the globe, and therefore, the World Social Situation Report, 2016 has also highlighted the importance of identification of patterns of social exclusion and discusses whether development processes have been inclusive or not.

Policy Shapes Power Relations

Policies combined with institutional arrangement and processes are entwined to shape power relations between the State as duty bearer and peasants in specific women as rights holders. Since peasants in specific are being excluded from the mainstream process of broad based inclusive growth for many decades, they will need to be supported to understand their rights and entitlements, be aware of the processes and articulate their voices in a meaningful way. In relation to peasants, agriculture is one of the major sectors that directly influence their entitlements and rights. Below is the brief presentation of agricultural policies, which have peasant and women specific provisions as presented in the Table 1.

Table1:Major Policies and Programmes in Agriculture Sector with Peasant and Women Specific Provisions

1. a Peasant Specific Objectives/Provisions

Major Policies in Agriculture Sector in Nepal	Preamble/Main Objectives	Peasant Responsive Objectives/ Provisions
Constitution of Nepal, 2015	Article 10.3: Right to equality Article 42.1 Right to social justice 42.3 Article 51.13.e Policies relating to land reforms	Nothing to prevent to make law for special provisions for ….farmers… ….farmers…shall have right to participate in the state bodies… shall have the rights to access to the lands for agricultural activities, protection of seeds… in accordance with the law ..land reforms in regard to the interests of farmers… land management, commercialization, industrialization…while protecting and promoting rights and interests of farmers farmer's access to agro-inputs, fair prices, and market

Aagriculture Perspective Plan- APP (2052)	Impact: poverty and food security	Organize rural poor people in administering food security programme and participation in broad growth process Employment opportunities for landless and chronically resource-poor Facilitate access to land
National Agricultural Policy (2061)	Access to land Policy framework	Identification of resource-poor as having below 4 hectares of land Faciliiate access to land for skilled labor Land bank Research and development Food deficit districts to have provisions for increment in productivity and or facilitate supplies Capacity building of extension workers and farmers Participation of stakeholders in implementation and monitoring
Agro-business Promotion Policy (2063)	Policy implementation, coordination and monitoring	Committee will be formed (no specific reference to peasants or farmers)
Food Act, 1966		No specific reference to peasants/ farmers
Consumer Protection Act, 1998	Preamble Consumer Protection Council	Protection of consumers No specific measures in reference to peasants/farmers as producers, traders, processors. All are dealt with under the purview of consumer

Agricultural Development Strategy (2015-2035)	General Overview Vision Strategic Framework ADS framework: improved governance	Farmers considered as one of the main stakeholders Inclusive % of farmers reached by agriculture programmes Recognized socio-economic diversity of farmers and categorization of farmers into landless, subsistence farmers and commercial farmers Provision for strengthened farmers' rights Refers to capacity of government to design, implement policies and discharge function for accountability, participation, predictability and transparency Inclusion taking into account through appropriate mechanisms. However, farmers' participation in decision-making bodies are explicitly spelled out

1. b Women Specific Objectives/Provisions

Major Policies in Agriculture Sector in Nepal	Preamble/Main Objectives	Women Responsive Objectives/Provisions
Constitution of Nepal, 2015	38.4 Rights of women Schedule 6.20 Schedule 8.15	Women shall have the right to participate in all bodies of the state on the basis of the principle of proportional inclusion Agriculture and livestock development, factories, industrialization, trade, business, and transportation Agriculture and animal husbandry, agro-products management, animal health, and cooperatives
APP (1996)	Transform subsistence agriculture to commercial	Improve women's participation to enhance agricultural productivity
National Agricultural policy (2061)	Transform subsistence agriculture to professional and competitive system	50% involvement and participation of women in all programs as possible

Agro-business Promotion Policy (2063)	Support for commercial and competitive agriculture for internal market and export	Special programme to support enterprises established and promoted by women, *Dalits*, and poor
Food Act, 1966		No specific reference to women, girls and socially excluded groups
Consumer Protection Act, 1998	Consumer Protection Council	Provision for two women representatives nominated by the government from among women working in fields connected with the rights and interests of consumers as members, No other specific measures in reference to women, girls and socially excluded groups. All are dealt with under the purview of consumer
Agricultural Development Strategy (2015-2035)	General overview Social and geographic inclusion Targeted interventions and indicators of inclusivity; food and nutrition	Recognized dominance of women farmers in the agrarian structure of Nepal Budget allocations are often silent over how to enhance women's strategic positions through recognizing women as independent and autonomous farmers, ensuring women's access to means of production, enhancing their leadership competence and creating acceptance, and improving women's position in different structures of the government, non-government and private sectors. Gender-balanced approach that explicitly ensures the participation of women % of farm land owned by women or joint ownership, Percent of rural households covered by agricultural services and programs, Enhancement of qualitative and quantitative aspects of participation of men and women farmers from all gender and social groups, Raise awareness on women's rights to land, Build capacity of women farmers in irrigated agriculture and water resource management, Agro-entrepreneurship programs for youth, women, disadvantaged groups and regions

Source: Compiled by author from different sources in 2016/17.

The Table above shows that Nepal has established quite a number of strategic documents related to poverty reduction and agricultural development. However, the documents do not necessarily deliver inclusive results for the benefits of peasants. Among those, agriculture specific documents are more relevant to peasants and women as they are the main custodians related to the sector. To translate the policy provisions into implementation, the Government of Nepal has been preparing periodic plan since 50's. The Table below shows how women responsive provisions are made in the periodic plans of Nepal so far.

Table 2: Progression of Women Inclusion in Periodic Development Plans

5th Periodic Plan (1975-80) -	establishment of women's service coordinating committee with inclusion of women focused programmes in different sectors including agriculture
6th Periodic Plan (1980-85) -	conceptualization of women inclusion in development programs with emphasis on meaningful participation of women
7th Periodic Plan (1985-90) -	active participation of women and quota for women, ensuring at least 10 percent of women participation
8th Periodic Plan (1992-97) -	women specific sub-sector programme such as group formation, training with the focus on institutionalization of inclusion provisions
9th Periodic Plan (1998-02) -	women in decision-making, and post-harvesting programs
10th Periodic Plan (2002-07)-	gender mainstreaming through capacity building and entrepreneurship
Three Years Interim Plan (2007-10) -	inclusive development and targeted programs; women contribute 48.9 percent of the labor force managing agriculture and other economic activities; priorities for employment to women
An Approach Paper to the 13th Plan (2013 -16) -	participation of women in targeted programmes, involve women in the development process and ensure they benefit from the results, improve access to employment opportunities, exemption in land registration fees, establish women entrepreneur's development fund, promote gender equality and women's empowerment, strengthen role of women in economic, social, and political empowerment

Source: Compiled by the author from different sources, 2017.

The international frameworks and procedures directly and or indirectly influence these national efforts. Nepal is signatory of those instruments and has obligations to fulfill those requirements legally if not morally as well. Some of the major international instruments and its provision in specific to economic, social, cultural rights are presented below in Table 3.

Table 3: List of International Instruments Ratified
by the Government of Nepal

S. No.	Instruments	Date of Signing/ Ratification	Major Provisions
1	Universal Declaration on Human Rights (UDHR), 1948		Article 25: Right to food, clothing and shelter for all
2	International Covenant on Civil and Political Rights, 16 December 1966	14 May 1991 a	Art. 1: ...pursue their economic, social and cultural development, Art. 6; Right to life
3	International Covenant on Economic, Social and Cultural Rights, 16 December 1966	14 May 1991 a	Art. 1: ...pursue their economic, social and cultural development; Art. 2.1 and 3: .... full realization of the rights and guarantee economic rights, Art. 7: employment and fair wages, Art. 9: social security, Art. 11: adequate food, clothing and housing
4	Convention on Elimination of all Forms of Discrimination against Women (CEDAW),	22 April, 1991	Art. 14g: equal treatment in land and agrarian reform as well as in land resettlement schemes; Art. 15: ... contracts and to administer property
5	Convention of Biological Diversity (CBD), 1992	15 September 1993	Local community rights over biodiversity, its use and benefit sharing
6	World Trade Organization (WTO), 1995	23 April 2004	Article 27.3b of TRIPs: patent rights
7	Convention on Child Rights (CRC), 1966	14 September 1990	Art. 24c; provision of adequate nutritious foods and clean drinking water

8	International Treaty on Plant Genetic Resources for Food and Agriculture (ITPGRFA)	2nd January 2007	Art.1: fair and equitable sharing of the benefits arising out of their use; Art. 5: exploration, conservation and sustainable use of plant genetic resources for food and agriculture; Art. 6: sustainable use of plant genetic resources; Art. 9: farmer's rights: protection of traditional knowledge, right to equitably participate in sharing benefits and right to participate in making decisions
9	International Labor Organization (ILO) 169*What are the RtF and FS provisions contained in the regulatory frameworks with specific reference to the Constitution 2015, Food (3rd amendment) Act, 1992 and Consumer Protection Act, 2054 (1998)?*	22 of August 2007 Literature reviews and interview with the Parliament, National Planning Commission members, Ministry of Finance, MoAD and MoFALD authorities	Rights of local and indigenous communities over natural resources e.g. Art. 14.1: The rights of ownership and possession over territorial land; Art. 15.1: The rights of the peoples concerned to the natural resources, right to participate in the use, management and conservation; Art. 17.1: transmission of land, etc.

Source: Ghale Y. and Bishwakarma, N. 2008

The tables above show that, there have been conscious efforts from the State towards inclusive development of the country in specific to agriculture sector for a long time. In terms of numerical achievement, Nepal had made substantive progress.

The report on progress of women in Nepal (1995-2015) also claims that there has been some key achievements in securing women's rights singe the Beijing fourth world conference on women as shown below in the Figure 2.

A more gender-friendly constitutional and legal framework	A substantial improvement in women's representation in the public sphere	Increased access to education and higher educational achievement	Significant improvements in maternal and child health	Increased access to economic resources, land, property and micro-credit

Figure 2: Key Achievements since Beijing

Source: Sahabhagi, Didi Bahini and FEDO, 2015

Likewise, Government of Nepal also took initiative in sharing the information on entitlements specific to women, which is important to be aware of right to information provisions. However, Nepal still has a lot of scope to excel in achieving peasant and women's empowerment with proper implementation of those policy provisions. The chapter below explains how the policy processes are linked with analytical frameworks and how they influence the governance system.

Analytical Framework: Determination of the Governance by the Policy

To begin with, Agriculture Perspective Plan (APP) 1996-2015 is taken as an example for the analysis of public policies and its provisions for inclusive results. The document shows inputs and outputs relations and the impacts it has yielded through the governance. The document as such does not have specific objectives to support peasants and women to be part of the transformational process for their empowerment except the gender mainstreaming impact objective as depicted below in Figure 3.

Inputs	Outputs	Impacts
• fertilizers • irrigation • roads • technology • credit • rural electrification	• livestock • high value commodity introduction • agribusiness • forestry	• poverty reduction • food security • environment management • gender mainstreaming

Figure 3: Major Foci of the APP, 1996

To continue framework analysis for inclusive mechanisms that exist in Nepal, it is important to understand available inclusive provisions and mechanisms. Table 3 below provides an overview of institutional mechanisms for Gender Equality and Social Inclusion (GESI).

Table 4: Institutional Mechanisms for GESI of the Government of Nepal

Level	GESI Mechanism
Central	National Planning Commission; Ministry of Women, Children and Social Welfare (MWCSW) and its Department of Women Development; Ministry of Federal Affairs and Local Development (MoFALD) and its *Dalit* and *Adibasi/Janajati* Coordination Committees; constitutionally established National Commissions for Women, *Dalits*, Indigenous Nationalities, *Madhesis, Muslims, Tharus* and a National Inclusion Commission that is mandated to protect the rights of *Khas Aryas, Pichardiaka* ("backward) class, persons with disabilities, senior citizens, labourers, peasants, minority and marginalized communities, backward classes, people of Karnali and the indigent class;[1] Gender/GESI Focal Points (GFPs) in NPC, MoFALD, MWCSW and the Ministries of Education, Health, Urban Development, Forestry and Agriculture.
District	Women and Children Offices (WCOs), Social Committee with a Social Development Officer of District Development Committees (DDCs); Indigenous Ethnic District Coordination Committee and *Dalit* Class Upliftment District Coordination Committee, the Gender Mainstreaming Coordination Committee (GMCC), and the GESI Implementation Committee
VDC/ Municipality	Representative Integrated Planning Committees (IPCs) in each VDC; Ward Citizens' Fora and Citizen Awareness Centres (CACs)

Source: IDPG, 2017

---

[1] This list of National Commissions and the groups covered is taken from Section 259 (a) of the constitution.

To understand the gaps in policies, its operationalization and exploit the possibilities further, there is a proposal made by International Development Partner Group (IDPG) as shown in Figure 4 below.

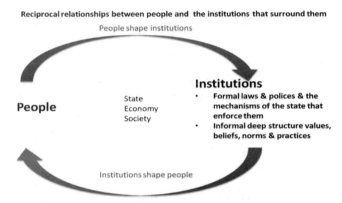

Figure 4: From Exclusion to Inclusion: People Shape Institutions; Institutions Shape People

While making such analysis, it is important to understand who are counted, who gets space and whose voices get heard and are able to create an influence. The Figure 5 below shows how women in general and other minority groups are involved in the policy processes.

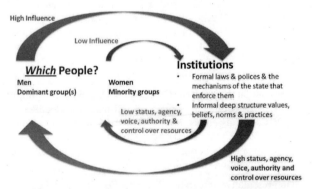

Figure 5: From Exclusion to Inclusion: Influence of the Type of People on the Institutions Shaping the State, Economy and Society

Source: IDPG, 2017

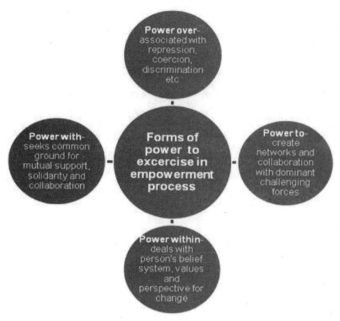

Figure 6: Different Forms of Power

It clearly shows that among all, women are less influential in the public policy processes. In claiming their rights and being a part of the policy processes, it is of utmost importance to make investment in capacitating them in understanding the forms of power as explained in Figure 6. The chapter below shows the situation in reality in transforming policies into practice.

The Ground Reality: Some Cases of Peasants' Situation in Nepal

Case 1. Denial of Access to Livelihood Resources for Sonaha

About 300 Sonahas of 42 households live in Saijana village of Manau Village Development Committee of Bardiya district. They represent one of the most endangered indigenous groups in Nepal. They lead semi-nomadic lifestyle and their livelihood mainly depends on fishing and gold panning. Since the establishment of Bardiya National Park in 1975, the livelihood of these Sonahas has significantly been weakened and as a result, they do not have free and unhindered access, as they used to enjoy in the past, to fishing, gold panning, and forest resources.

The stricter regulations and limited access to the park and river have drastically reduced the earning of these Sonahas. Most Sonaha women who were skilled at gold panning were forced to abandon their original occupations, without any alternative arrangement in place. While establishing the national park, no consultation was done with Sonahas. The strict regulations imposed on this and other communities living in and around the national park have but negatively affected in their capacity to accessing food.

Case 2: Suffering of Balapur Inhabitants

The right to food of 1377 Balapur villagers in Banke District is at risk after the establishment of Sikta Irrigation Project and Banke National Park in May 2010. The village located at the buffer zone 120 of Banke National Park is surrounded by forests, covering an area of about 200 *bigha* (6.6 hectares). Forest resources are the main source of livelihood for the villagers. Farming lands, houses, livestock and grain storages are at risk of being encroached by wild animals. Even their life is under threat from wild animals as the national park, according to the Government, is particularly set up to conserve the tigers.

In addition, the villagers' lands also face the risk of inundation due to Sikta irrigation project, especially during monsoon. The Balapur residents demand for alternative arrangements with appropriate resettlement in a suitable location and assurance of access to their sources of income and proper food. At present, no provision of alternative settlement was offered or the Government has provided compensation of other means of livelihood.

No adequate and proper consultation with the villagers and the community forest users' group has taken place prior to this decision. The villagers organized themselves, met local and central level authorities and submitted petitions with their demands which, among others, included either resettlement to another areas or protection of their life and crops from wild animals as well as unhindered access to forest and resources to sustain their livelihood. On 12 February 2012, a peaceful demonstration of the villagers was even met with the excessive use of force by the district administration, leaving 42 people injured.

Case 3: Landless Gandharva Women

Lack of proper wages, social security and health care has been leading to severe food insecurity and malnutrition of women in Surkhet district. In a hamlet composed out of small huts without doors or windows live 14 landless women aged 22 to 45, belonging to the Gandharva community. They suffer from severe lack of food and basic health care facilities.

These women have lived in Jhupra hamlet, Baluwatar 1, of Jarbutta VDC since 1993, located in the Shiva community forest between two other community forests. Their daily work entails collecting sand from the Jhupra stream banks, crushing stone and loading the heavy stones into trucks that transport these to the cities, where the material is used for construction work. For this, they hardly get paid 100 NRs (around USD 1) a day by the middleman for work done from dawn to dusk.

Nutritionally, these women hardly get to eat two meals a day and most of the time they have to satisfy themselves with simple rice taken with "*fado*", a mixture of water and wheat flour.

This gets even more severe during the pre-natal and postnatal conditions with lack of diet and detrimental effects on the health and nutrition of both mother and child. There is a lack of protein intake (i.e. milk, egg, fish) during pregnancy. Women have to work during their entire pregnancies and also immediately after giving birth. Immediately, these women do not have any alternative or other income source to change their situation. Hunger and malnutrition exacerbate their health problems day by day. Their earning, which is below par the minimum wage per day determined by the local government, is just insufficient to manage even daily food stuff of 4-8 family members. This has made them compromise with the enjoyment of their other basic rights, including education to their children (Source: Fian Nepal, 2014).

Case 4: Women in Cooperative Management

To give an example, Nepal government has taken the **three pillars approach of economic growth**, namely, public, private and cooperative. Most of the cooperatives, especially in remote areas are operated with high participation of women. In addition, most of those women operated and/or with majority women shareholders are focused in the agricultural sector related cooperatives. However, the investments in those cooperatives are almost negligible. The table below depicts the scenario in cooperative management.

Table 5: Women Participation in Cooperative Management

Executive members		General members		Staff	
Female	Male	Female	Male	Female	Male
91,196	1,76,526	2,67,988	22,81,935	26,886	31,191

Moreover, the investment made in the cooperatives with the involvement of large number of women are engaged is quite low. Though the targets are largely met, it is yet to address the structural causes of discrimination, exclusion and powerlessness.

All the above cases show that there is systematic exclusion of peasants, in particular the women peasants, from the policy processes and their operationalizations. There are gradual improvements in the inclusive policy recourse; however, the systematic causes of exclusion and deprivation are yet to be responded by the state. While ensuring good governance with the aim to enhance accountability, there is someone to be held accountable. In this case, the state as duty bearer has the responsibility to ensure rights to its citizens as rights holders through mobilization of different actors as responsibility bearers. Human rights create rights and entitlements to one party and corroborative duties to the other. State is the ultimate duty bearer, accountable to ensure the fulfillment of the rights of the rights holders.

Case 5: Policy Shortcomings

> The survey for the first time sought information on women's participation in fifteen types of different household decisions: children's education, choice of school for children, personal (her own) health care, pre-natal care, use and method of family planning, children's health care, expenditure on food, other household expenditure, selling household goods (including livestock), crop cultivation, receiving credit, use of loans, leaving home for job, and use of remittances received. The questions were asked to spouse of male household head (Section 9D) or the female household head (Section 15D).

Source: NLSS, 2011

The box above explains that after such a long time, women's participation in decision-making has been assessed. However, all the indicators are focused on transactional spheres. Nowhere it has mentioned about these spaces as being taken as entry points for transformational changes. This is one of the best examples of policy flaw which shows lack of inter-linkages with the policy frameworks and assessment procedures, which are to be further linked with programmes and plans accompanied by appropriate institutional arrangements and resource allocations. To deal with the public policies and peasantry in relation to poverty reduction, inclusion and empowerment of peasants, more specifically with women peasants, it is important to understand the fundamental concepts of food security, right to food and food sovereignty. It will then help design governance mechanisms that can seek people's participation, representation and influence decision-making processes.

Understanding Concepts on Food

Food Security

The most commonly used definition of food Security as defined by FAO during the World Food Summit in 1996 is *"Food security exists when all people, at all times, have physical and economic access to sufficient safe and nutritious food that meets their dietary needs and food preferences for an active and healthy life."* (FAO, 2008).

The main pillars of food security are explained as:

Availability   availability primarily deals with the 'supply side' of food security which is largely determined by food production, stock, trade, and distribution

Accessibility   Accessibility denotes for either physical access and or supplies during emergencies and disasters through social safety provisions. It is, therefore, important to assure that physically available food items are accessible by all people at the time they require in a fair price, of required amount and of cultural acceptance

Affordability   Food physically available at certain time and places does not always mean that all people have the purchasing capacity. Therefore, the creation of a conducive policy environment that facilitates for employment, incomes, expenditure, markets and prices determine the affording capacity of the people and determine food security situation is crucial.

Stability   One of the major areas of responsibility of the state is to ensure regular supply of food even in adverse weather conditions, political instability, or economic upheavals with specific responsibility to the vulnerable group of people

Utilization   Utilization stands for both biological utilization of available food as well as ensuring the nutritional status of food during preparation, feeding practices with special attention to the diversity of the diet and intra-household distribution practices

Food security by definition, therefore, mainly deals with the technical aspects of food production, distribution and supplies. However, it is equally important to understand that food is vital to human survival. Therefore, right to food is right to life. Against this backdrop, the concept of Right to Food is explained below.

Right to Food

The concept of Right to Food (RtF) was introduced by American President Roosevelt within the framework of Free from want in 1941 (Franklin D. Roosevelt's Address to Congress). The concept is further taken up by International Instruments such as Universal Declaration of Human Rights (UDHR), 1948.

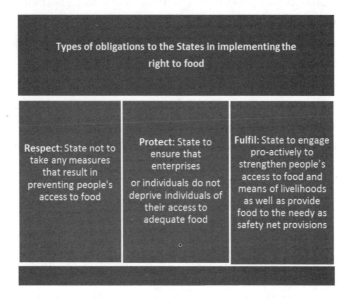

These instruments provide an important basis to understand food as basic human rights with corroborative obligations of the state to ensure progressive realization of right to food to its people. In this juncture, Food and Agricultural Organization (FAO) has developed the Voluntary Guidelines on Right to facilitate the member states and other actors to develop common understanding of the provisions and translating it into the domestic policy frameworks, (FAO, 2004). The framework mainly refers to three major pillars namely, respect; fulfill and protect.

Since, the provision for right to food is enshrined in the International Human Rights Framework documents, individuals as *rights-holders* can make legitimate claims, and States as *duty-bearers* along with the support of other actors as *responsibility bearers* can be held accountable for their acts or omissions in the policy frameworks and institutional arrangements.

Therefore, proper articulation of functions, supported by adequately capacitated functionaries and proper allocation of funds is quite crucial in fulfillment of the right to food to all people.

Though the right to food is a normative obligation for all the nation states as a member of these human rights frameworks, its proper implementation is still not adequate. The situation is further compounded by the market-led processes to deal with food and its total chain, and the tendency of decision-making at the global forums wherein economically weak nation states have less power to influence the decisions. However, the peasant movements such as La Via Campesina and other like-minded groups around the globe have been advocating the food sovereignty to retain state power to decide its food regime. This implies that managing food system is a political process rather than a mere technical and or normative framework based human rights principles. The chapter below explains what food sovereignty deals with.

Food Sovereignty

Food sovereignty is an emerging concept mainly being promoted by groups who are closely associated with the nature, environment, and peasant's concerns. They believe that the neo-liberal economies promoted through multilateral trade forums such as World Trade Organization (WTO) are creating more challenges and negative impact on sovereignty of the nation states especially of the south.

The definition of Food Sovereignty was actually coined by Via Campesina, a peasant's movement group, as *"Food sovereignty is the right of each nation to maintain and develop its own capacity to produce its basic foods respecting cultural and productive diversity. We have the right to produce our own food in our own territory. Food sovereignty is a precondition to genuine food security."* (Lee, 2007 adapted from Via Campesina, 1996, p.1).

Moreover, scholars such as Pimbert (2009, p.4) claims that moving towards food sovereignty is about reclaiming autonomous food system by the nation states. He argues that it is a transformative process which aims to guarantee and protect people's space, ability and right to define their "own" models of food production, distribution and consumption patterns. It, therefore, recreates democratic realms and regenerates diversified autonomous food system that leads to equity, social justice and ecological sustainability. The concept, therefore, demands a holistic and broader framework that deals with four inter-related aspects as follows: (i) political; (ii) economic; (iii) social, and (iv) ecological. He further elaborates that localized food systems help generate jobs and employment opportunities mainly for smallholders and small and medium-size enterprises which are resilient for the country's economy towards vulnerability.

In this pretext of understanding food security, right to food and food sovereignty, it is crucial to define and establish a sound food security governance system. The context clearly reveals that each concept has different historical connotation and the scope. However, each of the concepts is evolving, dynamic and mutually inclusive. The food security governance system, therefore, should aim to ensure the mechanisms for proper state-people interface to make the state accountable to its people for progressive realization of right to food. The chapter below explains the food security governance and its principle.

There is no internationally accepted definition of the food security governance. However, the generic principles of good governance apply to the food security governance as well. For the purpose of this research, food security governance is dealt with in reference to the specific context, and governed by the system throughout the chain of production, distribution, supplies and access. In general, governance of food system refers to the actions as follows: (i) creating and exchanging food security-related information (policies, decisions, practices etc.); (ii) setting standards and actively promoting the knowledge required to achieve food security goals; (iii) ensuring coherence between food security strategies, policies, rules and regulations; (iv) providing technical assistance and setting up food security projects and programmes that cut across borders, and (v) coordinating international assistance during crises, disasters and food security emergencies.

Definitely, there has to be the improved global governance for hunger reduction programme, and ensuring the right to food. In this context, the major principles of the food security governance are explained below.

Accountability: Accountability is the key tenet of good governance for promoting rule of law and transparency while implementing the programme. The state has to establish rules and decisions through participatory and inclusive processes for accountable outcomes. The state has to demonstrate that it has properly followed good governance procedures, and used tools for monitoring and evaluations. Indeed, accountability is one of the major principles of human rights to empower people as well as ensure responding public policies are in place.

Participation: It refers to the participation of diverse group of stakeholders including citizens, specially vulnerable group's either directly or through legitimate representatives. There have to be the mechanisms to ensure informed participation, freedom of expression and best interests of the people. There has to be the inclusion of the stakeholders in the policies, processes and institutional set ups.

Transparency: It pertains to the sharing of information in easily understandable language. It is to be freely available to those who are directly affected by those decisions and policies. There has to be the mechanism to ensure whether the enforcement is in compliance with the set rules and procedures or not.

Efficiency and Effectiveness: They pertain to: (i) best use of resources at people's disposal; (ii) sustainable use of natural resources, and (iii) protection of environment. Making difference over a period of time is equally important in the domain of effectiveness.

Rule of Law: Under it, fair legal frameworks have to be enforced impartially and there has to be the full protection of human rights.

Complying with the right to social security, the state must ensure that social protection is *equally available to all* individuals, and in this respect, direct their attention to ensuring *universal coverage, reasonable, proportionate and transparent eligibility criteria; affordability* and physical *accessibility* by beneficiaries; and *participation* in and *information* about the provision of benefits.

Food Sovereignty

La Via Campesina has adopted seven principles of food sovereignty which are presented below.

(i) Food: A Basic Human Right

Everyone must have access to safe, nutritious and culturally appropriate food in sufficient quantity and quality to sustain a healthy life with full human dignity. Each nation should declare that access to food is a constitutional right and guarantee the development of the primary sector to ensure the concrete realization of this fundamental right.

(ii) Agrarian Reform

A genuine agrarian reform is necessary which gives landless and farming people –especially women – ownership and control of the land they work and returns territories to indigenous peoples. The right to land must be free of discrimination on the basis of gender, religion, race, social class or ideology; the land belongs to those who work it.

(iii) Protecting Natural Resources

Food Sovereignty entails the sustainable care and use of natural resources, especially land, water, and seeds and livestock breeds. The people who work in the land must have the right to practice sustainable management of natural resources and conserve biodiversity free of restrictive intellectual property rights. This can only be done from a sound economic basis with security of tenure, healthy soils and reduced use of agrochemicals.

(iv) Reorganizing Food Trade

Food is first and foremost a source of nutrition and only secondarily an item of trade. National agricultural policies must prioritize production for domestic consumption and food self-sufficiency. Food imports must not displace local production nor depress prices.

(v) Ending the Globalization of Hunger

'Food sovereignty' is undermined by multilateral institutions and by speculative capital. The growing control of multinational corporations over agricultural policies has been facilitated by the economic policies of multilateral organizations such as the World Trade Organization (WTO), World Bank (WB) and International Monetary Fund (IMF). Regulation and taxation of speculative capital and a strictly enforced code of conduct is therefore needed for ensuring food sovereignity.

(vi) Social Peace

Everyone has the right to be free from violence. Food must not be used as a weapon. Increasing levels of poverty and marginalization in the countryside, along with the growing oppression of ethnic minorities and indigenous populations, aggravate situations of injustice and hopelessness. The ongoing displacement, forced urbanization, repression and increasing incidence of racism of smallholder farmers cannot be tolerated.

(v) Democratic control

Smallholder farmers must have direct input into formulating agricultural policies at all levels. The United Nations and related organizations will have to undergo a process of democratization to enable this to become a reality. Everyone has the right to honest, accurate information and open and democratic decision-making. These rights form the basis of good governance, accountability and equal participation in economic, political and social life, free from all forms of discrimination. Rural women, in particular, must be granted direct and active decision making on food and rural issues.

Role of Food Security Governance for Ensuring Rights of Peasants

Let us understand some definitions of governance. WB defines governance as "the way...power is exercised through a country's economic, political and social institutions". Likewise, UNDP defines governance as "The exercise of economic, political and administrative authority to manage a country's affairs at all levels. It comprises of mechanisms, processes, and institutions, through which citizens and groups articulate their interests, exercise their legal rights, meet their obligations, and mediate their differences".

Moreover, Duncan J. (2011) defines governance as "a process of governing. It is the way in which society is managed and how the competing priorities and interests of different groups are reconciled.It includes the formal institutions of government but also informal arrangements. Governance is concerned with the processes by which citizens participate in decision-making, how government is accountable to its citizens and how society obliges its members to observe its rules and laws. Governance comprises the mechanisms and processes for citizens and groups to articulate their interests, mediate their differences, and exercise their legal rights and obligations. It is the rule, institution, and practice that sets limits and provides incentives for individuals, organizations and firms."

It clearly shows that governance is a principle based framework with its strong focus on the accountability result, which encompasses all economic, political and social spheres. Governance in principle encompasses five common pillars as follows: (i) accountability; (ii) participation; (iii) transparency; (iv) efficiency and effectiveness, and (v) rule of law.

The above analysis shows the realization of good governance into practice is quite challenging. Referring to food governance, it demands the state as duty-bearer to exert its power for the discharge of its responsibility with full competence to respect, fulfill and protect people's rights to food. Likewise, people as rights-holders have certain roles and responsibilities while claiming and enjoying their right to food. Some of the major ways the government can play (Cargill, 2014, p. 1) its role are mentioned as follows: (i) investing in agriculture; (ii) supporting small-holder farmers; (iii) harmonizing food safety standards; (iv) enabling market space; (v) reducing environmental impacts; (vi) balancing production and trade, and (viii) facilitating emergency food aid.

In connection to above discussion on food security, rights to food and food sovereignty in reference to food governance, it is worth mentioning that the Agriculture Development Strategy (ADS) has considered governance as one of the major pillars of the strategy as shown below in Figure 8.

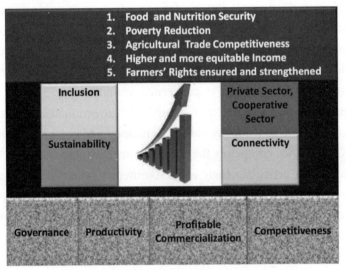

Figure 8: Major Characteristics of the ADS

Source: GoN, 2015.

Conclusions: Need of Proper Assurance of Empowerment of Rights Holders and Accountability of the State

The above analysis confirms that there is a need of proper understanding of the concepts of food security-related frameworks, and state's obligations and mechanisms to ensure rights holders, particularly women's and girl's empowerment to claim their rights responsibly. It requires proper investment plans in both targeted and mainstreaming programmes for holistic improvement of governance. Therefore, two-pronged approach with the balanced focus on empowerment of rights holders and accountability of the state would be the plausible way- forward.

Empowerment of the Rights Holders through Meaningful Participation

Since peasants, particularly women are being excluded from the access to and control over means of production, participation and representation in decision-making forums and influence the decisions to make their livelihood options more recognized and feasibly sustained, there is a need to make conscious effort and invest for their empowerment. While doing so, it is important to consider following four important stages systematically:

Engagement – ensure participation and representation of peasants in public policy processes to understand and claim their rights responsibly;

Empowerment – support to acquire ideas, knowledge, skills and resources to build their self-confidence;

Enhancement – facilitate the application of knowledge and skills to enhance life of families and the community they belong to, and

Emergence – mobilize them to move on to a public stage for social and political action that transforms environment.

Accountability of the State through Policy Frameworks and Institutional Arrangements

To shape the state's role in facilitating inclusive public policies, institutional frameworks and inclusive results, it is important to define functions, functionaries and funds with proper investment mechanisms and mandate to have receptive attitude to discharge its function. In the context of federal state building process, it is the opportune moment to ensure the constitutional rights being effectively implemented by different tiers of governments. The exclusive rights given to the local level in particular provided space to sub-national governments to remain accountable for meaningful and inclusive participation of local communities (i.e peasants and discriminated groups) in broad-based growth policy formulation, implementation and monitoring.

References

APROSC and John Mellor Associates Inc., (1995). *Nepal agriculture perspective plan, 1995/96-2014/15 A.D. (2052/53-2071/72 B.S.).* Final Report for National Planning Commission, Government of Nepal and Asian Development Bank, T.A. No. 1854-NEP, Kathmandu

Cargill (2014). *Important role of governments in achieving food security.*

CBS (2011). *Nepal living standard survey 2010/11.* Central Bureau of Statistics, National Planning Commission Secretariat, Government of Nepal, November 2011, Kathmandu.

Duncan J. (2011). *World food security policy and global governance: Towards fragmentation and arborescence, presentation of preliminary reflections on global governance.* London: Centre for Food Policy, City University, the UK .

FAO (1996). *Declaration of world food summit.* Rome: Italy.

FAO (2004). *Voluntary guidelines to support the progressive realization of the right to adequate food in the context of national food security.* Adopted by the 127th Session of the FAO Council, Rome, Italy.

FAO (2008) An introduction to the basic concepts of food security. *Food security for information action-practical guides.* The EC - FAO Food Security Programme, Rome, Italy.

FAO (2010). *Assessment of food security and nutrition security in Nepal* (An input for the preparation of NMTPF for FAO in Nepal), FAO, Pulchowk, Nepal.

Franklin, D. R. (1941). *The "four freedoms" address to congress.* Accessed on 06 January 2016: http://www. Wwnorton.com/college/history/ralph/workbook/ralprs36b.htm.

FIAN Nepal. (2014). *Parallel report on the right to adequate food in Nepal (Article 11, ICESCR)* FIAN, Kathmandu, Nepal.

Ghale Y. and B.R. Upreti (2008). Concentration and monopolization of seed market: Impact on food security and farmer's rights in mountains. *Mountain Forum* http://archive.mtnforum.org/rs/ol/browse.cfm?tp=aui&st=list2&authorID=2902&authorName=Yamuna%20Ghale.

Ghale, Y. (2016). *Gender equality campaign through cooperatives: Review and roadmap.* Paper Presented at the workshop organized by National Cooperative Federation on National Cooperative Day, 06 July 2016, Kathmandu.

Ghale, Y. (2010). Corporate globalization: Hunger and livelihood insecurity in Nepal. In B.R. Upreti and U. Muller-Boker (Eds.) *Livelihood Insecurity and Social Conflict in Nepal*. Kathmandu: Swiss National Centre of Competence in Research (NCCR), North-South

Government of Nepal (1966). *Food Act, 2052 B.S. (2066)*. Kathmandu.

Government of Nepal (1998). *Consumer Protection Act*. Kathmandu.

Government of Nepal (2006). *Agri-Business Promotion Policy,2063 B.S. (2006)*. Kathmandu.

Government of Nepal (2007). *APP Implementation Status Report, Volume 1 and Annexes*: APP Support Program (APPSP), Kathmandu.

Government of Nepal (2011): *Nepal living standard survey-III*. National Planning Commission Secretariat, Central Bureau of Statistics, Kathmandu.

Government of Nepal (2015). *Agriculture development strategy 2015-2035*. Kathmandu.

Government of Nepal (2061 B.S.). *National agriculture policy, 2004*. Kathmandu.

Hippel W. and Trivers R. (2011). The evolution and psychology of self-deception. *Behavioural and Brian Sciences* (2011) 34, 1-56, doi: 10.1017/ SO140525X10001354.

IDPG (2016). *A common framework for gender equality and social inclusion*, IDPG-GESI Working Group, Kathmandu.

MoAD (2073). *Gender equity and social inclusion annual progress report, fiscal year 2071/72 (2015/16)*. Ministry of Agricultural Development, Government of Nepal, Kathmandu.

Pimbert M. (2008). *Towards food sovereignty: Reclaiming autonomous food systems, reclaiming diversity and citizenship*. Rachel Carson Centre, Coventry University, The Centre for Ecology and Food Security.

Sahabhagi, Didi Bahini and FEDO (2015): *Progress of women in Nepal (1995-2015)*. UN WOMEN Nepal, Pulchowk.

UN (1979). *Convention on the elimination of all forms of discrimination against women*. Division for the Advancement of Women, Department of Economic and Social Welfare, United Nations, Geneva.

UN (2016) *Leaving No one behind: The imperative of social development*. Report on World Social Situation , Geneva.

Changing Labour Regimes of the Musahar Peasants: A Study from the Tarai of Nepal

MADHU GIRI

Introduction

This story begins with a peasant named Lalan Sada. He was born around 1955 A.D at the village of Jamdaha located in Siraha district of south-eastern *Tarai* of Nepal. It was sunny morning of October, 2012. A group *Musahar* youths from Bhangbari-JamAdaha were sitting on cemented *Chaputra*[1] under jackfruit at Bhangbari. The social life of Bhangbari *Chaputra* was mostly found occupied either for community meetings, informal discussions, experience sharing on migration or domestic issues of women, children, elderly and labour planning. Labour contractors and employers were found counting labour reserve in each household. Santosh

---

[1] The *Chaputra* was public place where the *Musahars* can take rest, midday nap as well as arrange public meetings. Though there were three *Chaputras* on the way from school gate to Jamdaha, Bhangbari *Chaputra* was predominated by the *Musaahrs* because it was located at *Musahari* village. School gate *Chaputra* was not only centre of the whole VDC but also resting place for firewood collectors, bazaar returnees, school intellectual community and any other visitors who pass through Jamdaha. Thapagau *Chaputra* was predominated by tea costumers and card players from its inception.

Thapamagar, local resident who has been teaching at Bardibas in Mahottari district, was waiting for *Musahar* labourer for paddy harvesting. He said that it was great mistake of planting paddy. He added," I will either plant mango on the field or leave barren next yeiar". I asked," why?" He replied that it was hard to find farm labourers in the community. He added," *Musahars* are not *Musahars* any more; they spend time by playing cards instead of working as farm labourers. At present, finding God is easier than finding a laboreurer". His statements were indicative of the dearth of farm labourers and negative attitude toward them. But they also hinted at the change of concept of labour in the community in the recent years.

In the morning of the same day, a group of *Musahar* youths were planning to go for beginning the wood cutting contract labour of Ramkumar Sah (Teli). They shared me that the land owner sold the wood to brick factory of Asanpur village. The manager of the factory contacted Ramkumar Sah (Teli) for wood cutting contract. Ramkumar Sah, the master contractor from Jamdaha, consulted with the *Musahar* youths at Musahari. Ramkumar gave wood cutting task to the youths on contract basis. According to Madhulal Sada, they agreed to cut five big trees by charging Rs.5000 (as per agreed sizes and pieces). They did not know how much money was charged by Ramkumar with factory manager. The woodcutter team added that the rate of cutting was *Sukha* (that meant they had to manage food and drinks by themselves). The woodcutter youths told their wives and family members to bring food and drinks on the wood cutting site which was about 20 minutes' walk from the settlement.

When we were talking about contract-based labour, an old unfamiliar man came and seated on the *Chaputra*. He told Lalan that he came to return his *Damposiya*[2] buffalo which was given by his wife and daughter without his consent. He was Jibchhi Sah (Sudhi) from Lalpur *Village Development Committee* (VDC). Lalan denied returning the buffalo. There was dispute between Jibchhi and Lalan. Both of them blamed each

---

[2] *Damposiya* is a system of livestock rearing contract or livestock sharing between owner and tenant.

other. Then, Jibachhi returned with gloomy face. Lalan told me the story of *Damposiya* buffalo that he has been rearing for the last five months.

Lalan Sada arranged two respectable marriages for his only daughter and third son in 2011 A.D. He was heavily indebted. Because of economic burden of two sequential marriage rituals, he had sold his milking buffalo. Then, he felt lack of buffalo at his residence. He was trying to find small *Damposiya* buffalo. He came to know that Jibachhi's wife was also interested to give her small buffalo as *Damposiya* because her daughter joined campus and she was separated with husband. Jibachhi has been living with another wife at next village and the buffalo was given to the first wife. Lalan and five other people went to the Jibachhi's wife and decided the original value of the buffalo (which was the sole ownership of Jibachhi's wife). Village notables decided the worth of Rs. 7500 as the original value and Lalan would get half of its value at the time of buffalo sale. Both parties agreed and Lalan took the buffalo at his home. Lalan's family had reared the buffalo. It was ready to be pregnant within a year. During that time, the market price of milking buffalo used to range from Rs. 50,000.00 to Rs.70,000.00.

Lalan also shared that his family has been sharing crop production with small landlord in the village. His family adopted multiple resource tapping strategies for the subsistence in the village. Like an experienced economist, Lalan had the answer here for diversifying labour portfolio to minimize household risk. He said, "The more the household has labour diversification, the less risk is for it from the volatility of agrarian production". I have observed that his daughter-in-laws collected firewood and sold at weekly markets. His first son practiced seasonal migration in India. The third son was planning to go Arabin country for employment. I have observed that Lalan's separated brother changed sharecropping land tenure into annual lease tenure and started vegetable production. The majority of the *Musahars* who practiced either sharecropping or lease land cultivation began contract off-farm labour and migration for their livelihood at Asanpur-Jamdaha.

By observing three stories above, I was interested to ascertain land-*Musahar* relations, changing labour regime from unfree agrarian labour to sharecropping and contractual labour of *Musahar* community. These two integrated forms of labor regimes have been operating simultaneously among the *Musahars*. Due to such change on labour regime, they depended on the most 'irregular livelihood' they had ever experienced. The irony of the free labour, then, is that everyone depends on market, but no one has what we are used to call a regular job or sense of livelihood security. My arguments are based on a limited understanding of free wage labour-cum-peasant livelihood, in which 'rational individuals' make decisions, considering the structural and material constraints. This paper attempts to demystify monolithic notion of land as private property resource and the *Musahars* were not first-rate cultivators. Therefore, on the contrary to the rhetoric of *Haruwa-Charuwas*, they were agrarian experts. I argued that they did not have much knowledge and inclination on agriculture *per se*. The first part of the paper deals with experiences and stories of *Haruwa-Charuwa* and its demise. The second section presents a discussion on sharecropping of livestock and crop farming and the last section demonstrates new dimensions of labour and livelihood among the *Musahars*. The paper is primarily based on fieldwork carried out for my Ph.D dissertation research at Golbazar-Jamdaha of Siraha district in 2011/012 and annual update visits were made between 2015 A.D and 2016 A.D.

Haruwa-Charuwa as Peasant Labourer

Alfred L. Kroeber (1948), who offers one of the best known early definitions of peasants, describes them as "part societies with part cultures." The *Musahars* (as a peasant class) existed as a part of the larger surrounding society, and retained cultural identities that set them apart. They were thought to hold relic of their cultural traditions: agrarian labour and more specifically *Haruwa-Charuwa*. The *Haruwa-Charuwa* is a colloquial term used in the central and eastern *Tarai* in Nepal to denote a particular type of agrarian labour. It clearly figured out unequal reciprocity particularly economic behaviors and generally political, social and cultural unfreedom. Lalan Sada, who became *Haruwa-Charuwa* in eight different landlord families for 40 years, shared certain degree of individual autonomy of landlord change and migration under the *Haruwa-Charuwa* system. They were employed by landlords for ploughing the land and herding the cattle. *Haruwas*, often senior males and heads of households, were males who ploughed land for other persons in exchange for either land or annual payment in cash or kind or to repay the debts. The term '*Haruwa*' refers to a type of agricultural labourer quite different from free labourer in agriculture. Under the free labour, a person works willingly for wages to satisfy his/her needs and can leave at any time for better opportunity. According to Jhokilal, one of the ex-*Haruwa* of Jamdaha, the *Haruwas* have *Laguwa* – an oral or written contract agreement that states that the worker would work on the landlord's land until the work would be completed. But they would not have the bonded status for all times. Mostly high caste, medium-sized and large landowners were employers of *Haruwa-Charuwa*. But in case of Jamdaha and some other northern villages of Siraha, *Tamang, Magar* and *Rai* households also employed *Haruwa-Charuwas*. *Jhokilal* and *Lachhana* had become *Haruwas* in *Tamang, Magar* and hill *Chhetri*. Their family members were also sometimes compelled to work for the same or different landlords. The *Laguwa* system was the same for other types of labour – *Charuwa*. The *Charuwas* were those people who were mainly employed for herding cattle. Mostly, the *Charuwas* could be children, women, elderly people or physically weak adults who cannot work as *Haruwas*, but work as cattle herders. The *Charuwas* were often the sons, daughters and elderly of *Haruwas*. A *Charuwa* could work either for a single or multiple landlords of the surrounding at the same time. Some time, *Charuwas* were similar to domestic servants living in the landlord's house.

A survey estimated that there were about 70,000 *Haruwa-Charuwa* households in the seven *Tarai* districts, accounting for 9 per cent of the total number of households (KC et.al., 2013). Two-thirds of *Haruwa Charuwa* households were *Tarai Dalits* and 95 percent were affected by forced labour system. The estimated total number of adults working under forced labour under the *Haruwa-Charuwa* system was 97,000, of whom a large majority (85%) were men (KC et.al., 2013). The reports made paradoxical argument. On the one hand, it said that *Haruwas* of

Central-East *Tarai* are forced/bonded labour, and simultaneously they are free to migrate to earn their livelihood during off-season on the other (KC et. al., 2013). According to Nebatiya Sada, about 25 percent of *Musahar* households were mobile either due to uxorilocality (Giri, 2017) or migration for work.Some of them migrated because of landlords' domination and possibility of unwanted forced labour. He added that some of them migrated to eastern part of Nepal and some of them went to India.

There is a lacuna of analysis on whether *Haruwa-Charuwas* could be viewed under bonded labour regime. According to ILO Convention No. 29 (1930), forced labour refers to 'all work or service which is exacted from any person under the menace of any penalty and for which the said person has not offered himself voluntarily' (ILO, 1930, Article 2(1)). It states that elements of forced labour are coercion, deception, exploitation and abuse, all of which deny human dignity, freedom and equality. On the basis of ILO definition of forced labour, all agricultural labour relations generally in South Asia and specifically in Nepal were put under forced labour. All forms of agrarian labour, mostly in *Madhes*, could not be homogenized as force labour category set by ILO in reference to European plantation labour. According to ex-*Haruwas'* stories, except debt bonded, most of *Haruwa-Charuwas* were/are free to change their landlords, dismiss the agreement after a year or migrate from one village to another village. The social life of the *Haruwa* system, like social life of things (Appadurai, 1986) had lives: birth, maturity, reproduction of variations and death. There was not only developed maturity of the system but also transformed on mode of payment from kind to cash. Though labourers had some degree of choice, they were not free from exploitation, deception and unequal/abusive treatment. My argument was that there was not homogeneous and static form of *Haruwa-Charuwa* in all times and places of *Madhes*. There was gradual evolution of *Haruwa-Charuwa* from forced to free labour. The later phase *Haruwa-Charuwa* labour relation was neither totally bonded because labour had multiple choices of livelihood nor totally free to choose type of work in particular place.

The *Kamaiya* Labour (Prohibition) Act 2002 stated. "The bonded labour system in Nepal is not exclusively limited within the *Kamaiya* system but is extended to other forms of labour in the agriculture sector. The *haliya, haruwa, hali, charuwa, bhunde* and *chakari* are other known forms of labour practices in agriculture under which bondage prevails. Suresh Dhakal argued that besides *Kamaiya* system, there have been reports of bonded labour elsewhere in the agricultural sector in Nepal. The *Haruwa-Charuwa* in central and eastern *Tarai* and *Haliya* in far-western hills/mountain were reported to be unfree agricultural labourers (Dhakal, 2007). He claimed that they were unfree, or bond(ed) labourers whose bargaining power was virtually non- existent, or has been surrendered. The unequal reciprocity between landlord and the *Haruwa* (Dhakal, 2007) was better analyzed in terms of politics of value (Appadurai, 1986). Very few *Haruwas* were there but the mode labour and politics of value have been changed recently in *Madhes*. My argument is that the classical 'Haruwa-Charuwa as a system' has collapsed because of labour out-migration, contractual system of labour and disappointment of farm labourers.

Jhoki Lal Sada told that *Haruwas* were employed on the day of *Sripanchami* - a Hindu religious day in which the Goddess of Education is worshipped in mid-February and is the day from which the farming season is supposed to start. His father was *Haruwa* and mother including himself were *Charuwas* at a *Magar's* household. When he was around 12 years old, father died and mother became helpless. Landlord used to visit his family on *Shreepanchami* asking whether Jhokilal could continue working as a *Haruwa* of the family. His mother assured that her family could plough the land. Once, at the time of paddy plantation, he got sick and could not plough land. Then, his mother was fined because of violation of agreement. In winter, he used to migrate to India and return at the very beginning of monsoon. Next year, his mother did not agree on *Shreepanchami* day; rather she decided during monsoon. Since then, he became *Haruwa* for long time. Though he was free to choose to become *Haruwa* or migrant labourer, he preferred to become *Haruwa* to stay with family members. Mostly, Haruwas were usually directly employed by landowners. There were few cases in which agents like *Munsi*, *Patawari* or Hatwe used to recruit *Haruwas*, when the employer owned a lot of land. This type of recruitment was found at Ghaletol where landlords rarely visited the land. These functionaries of *Jamindari* managed *Haruwa-Charuwa* recruitment, appointment and deletion. *Munsi*, *Patawari and Hatwe* were the 'white collar' employees of the *Jamidars*. *Munsi* supervised the work of *Haruwa-Charuwas* and other casual labourers. *Patawaris* was responsible to maintain records of loan disbursement and repayment. *Hatwe* was responsible for weighing the paddy to provide it to the labourers as wages or loan. These functionaries of *Jamindari* found defunct in *Madhes*. Jhokilal argued that matured form of *Haruwa*, in which landowner and tenant came to agreement after certain level of bargaining, prevailed in heartland of rural *Madhes*. Therefore, the relationships and nature of labour did not fit to call them *Haruwas* as the category understood in classical sense. There was not such classical *Jamindari* and landlordism. Medium-sized landowners were themselves in dilemma on whether to continue landownership or dissolve the land because of the dearth of farm labour. Most of the landowners were found happy if they found sharecroppers. In spite of domination and exploitation, both parties agreed that there was regularity and security of livelihood of landless peasants in *Haruwa* system. Land and farm embedded social relations have been found in crisis. There was confusion and contradiction among the *Musahars* for future mode of livelihood.

Interestingly, the *Musahars* were depicted as experts or knowledgeable agri-labourers (Jha, 1998, NNDSWO, 2006) but they were found that they had neither much knowledge nor interest on agriculture. Neither males nor females have much knowledge about maximum utilization of land and crop diversity. They were not first-rate cultivators like *Tharus* of the *Tarai*. When I observed barren land around Jhokilal's house, his wife replied that her family members had neither interest nor knowledge of green vegetable farming around the home. This was not only the case of *Jhokilal* and *Jamdaha Musahari*. I found paradoxical observation at *Ghaletol* and *Chuhade Musahari* where the *Musahars* of both settlements were employed on lease land by one NGO. None of the lease land cultivating candidates utilized land around their settlements. I have not observed single plant of vegetable or crop on spare land behind their huts. When I asked the *Musahars*, they replied that they had not much idea of vegetable, preservation of seeds nor they had much interest on utilization of the land. Nebatiya Sada, one of the leaders of *Ghaletol Lease Land Project* funded by WOREC Nepal[3], worked on 2 *Kattha* lease land but his half *Kattha* land around his hut was barren. He told me that his family never planted any crops and vegetables on the land. He added that he had neither seeds of crops and vegetables and tools nor he had much idea what to plant. Moreover, the *Musahars* told me that the idea of land as private property was contradictory to the *Musahars'* world view of land as common property. They told many cases of rejection to own land as private property. The rhetoric of *Musahars* as landless and expert agrarian labourers was used to develop the narratives to legitimize their landless labour status. The *Musahars* have different narratives regarding landownership and agriculture. They had collective resource ownership imagination and agriculture was not their main interest. Therefore, whole narrative of *Musahars* as landless and skilled agrarian labourers was ethnocentric predominated by *Jamidari* and capitalist world views.

Sharecropping and Labour in Transition

Lalan remembered the by-gone days when he was *Haruwa* and his family members were supporters of landlord family's jobs. Without the involvement of family labour, it was hard to run family economy and social life in the community. He told that besides small piece of *Jirayat* land (Regmi, 1972), there were few alternatives of livelihood. He added that the dense *Chure* forest was beneficial for hunting and livestock-raising, but firewood was not considered as a market commodity during those days. The main story of his life began when his parents died when he was less than 15 years old. He neither migrated to India in search of seasonal job nor he migrated to other parts of Nepal for long-term

---

[3] Women Rehabilitation Centre Nepal (WOREC) funded lease land project at Ghaletol in Siraha and Chuhade in Udayapur on the assumption that the *Musahars* became poor because of their landless status.

employment. He changed many landlords and experienced varieties of *Haruwa*. When political regime and bonded labour policies changed, he stopped working as *Haruwa-Charuwa*. At the same time, his elder son started to migrate to India for seasonal labour. His female members of the family had started to sell firewood at *Golbazar* and other emerging local markets (besides highway from *Lahan* to *Mirchaiya*). Then, remaining family members started sharecropping of small pieces of land, sharecropping of domestic animals and contractual labour. This subsistence economy was a common practice among the *Musahars* of Jamdaha. Though, *Haruwa-Charuwa* form of labour was declared illegal, few *Musahars* and other marginalized communities continued *Haruwa-Charuwa* in different forms. By showing a ploughman (a *Musahar* from *Musahaniya* settlement) in a *Tamang*'s farm, he said that the ploughman was *Haruwa*. But neither ploughman nor landowner named their agreement as *Haruwa*. The *Haruwa* was fixed on Rs. 20000.00 *per annum*. Lalan argued that the amount of money was very good in comparison to earlier years. Though many local people knew that but they did not disclose the secrecy of the villagers. Their household subsistence livelihood aptly matched what many anthropologists called 'peasantry'.

Eric Wolf (1966) and Marshall Sahlins (1960) distinguished peasants from primitive cultivators and agricultural entrepreneurs. They argued that in primitive society, producers not only control the means of production, including their labour and its product but also exchange surpluses directly among the members of the groups. Peasants however, are rural cultivators whose surpluses are transferred to a dominant group of rulers that uses the surpluses both to underwrite its own standard of living and to distribute the remainder to the groups in the society. Similarly, they differentiated peasants and capitalist farmers on the basis of location and mode of production. They said that peasants, as rural cultivators, raised crops and livestock in the countryside, not in greenhouse in the midst of cities or in aspidistra boxes on the windowsill. At the same time, 'peasants are not farmers or they do not operate an enterprise in the economic sense, they run a household, not a business concern' (Wolf,1966, Sahlins, 1960). The *Musahars* of Jamdaha sell firewood and vegetables at weekly markets, but the purpose of transaction was not accumulation of property as the spirit of capitalism. In her recent book *The Mushroom at the End of the World*, Anna L. Tsing calls this type of economic diversity as "salvage accumulation in pre-capitalist worlds"(Tsing, 2015,p.65). She added that through salvage accumulation, lives and products move back and forth between non-capitalist and capitalist forms; these forms shape each other and interpenetrate (p.65).

One of the powerful elements of the peasantry is sharecropping mode of production. Sharecropping is undoubtedly an extremely ancient method of agricultural exploitation in its origin and one which continues in many areas of the world down to the present day. Sharecropping is a system of leasing agricultural land in which rent is paid not as a fixed sum, but as a proportion of the harvest (Donaldson, 2000). The definition itself underlines the nature of the problem. When the contract is made, the crop is as yet unharvested, which is why sharecropping contracts fail to specify the date of payment and the exact value of the rent. Sharecropping can, therefore, be understood as involving risk and the potential for unjustified profit. In comparison to social life of bonded peasantry, sharecroppers were considered as uplifted free labour because the sharecroppers involved in dialogic social life and decision-making (Patterson, 1982). On the contrary, Gyan Prakash (1990) argued that social life of sharecroppers was not significantly different from bonded labour in North India. He argued that sharecroppers were not sovereign to use types of seed, fertilizer and time of cultivation. His case study of *Bhuiya* sharecroppers showed that sharecroppers were another form of bonded labour in terms of decision-making in the farm (Parkash, 1990). On the contrary, *Musahar* sharecroppers were observed relatively free from dictation of landowners. Shiblala Sada from *Jamdaha* has been cultivating a piece of one *Bigha* of land belonging to a *Tamang* family for the last five years. His landowner has not complained hitherto because the cultivation is done timely and there is no severe damage of crops. He added, "I do not prefer sharecropping because neither landowner nor cultivator gets satisfactory output. If I don't have *Damposiya* buffalo, and goats, I will decide to dismiss the sharecropping agreement. The *Tamang* community did not let our buffaloes and goats graze in their land without sharecropping or making the agrarian wage labour available in their field". He indicated sharecropping was embedded with livestock- rearing. In winter, some of rain-fed lands remained barren and these lands were grazing resources. Beside cereal crops, Shibalal got access to grazing lands and grasses grown there. He was not interested to receive the land on annual lease basis because he was not interested in farm labour. He shared," There is no certainty of agricultural output. If I go for contractual labour of construction and wood cutting, I will get satisfactory cash".

Damposiya was not sole cause that hooked them with sharecropping. He added, "There will be unintended consequences in the family like police case, crime and court affairs. The *Musahars* have no access to government offices and official resistance. Last year, his brother was arrested in the case of electricity line hooking. His landowner helped to release his brother from police case. The sharecropping relations could be useful resources to deal with adverse situation because landowners have wider socio-political network. In his line Vinayak Chaturvedi (2007) argued that peasants made non-economic and normative judgments by relying on practices and discourses already sanctioned by local culture ((Chaturvedi, 2007).

Though sharecropping was not attractive mode of production among the *Musahar* youths, their parents hooked their family labour on sharecropping. Senior *Musahars* argued that sharecropping was social phenomenon which connects between landowners and landless labourers. For them, sharecropping was not compulsion. Shiblal said that when paddy was harvested this year, landowners came into the field and requested to drop all his part of production at home. He added, "Landowner did not interrogate any aspect of production. He was happy because his field was cultivated. If I reject sharecropping agreement, he will make it mango garden. Some of his family members migrated and others are not interested in agriculture". For Shiblal, sharecropping was not only asset but also a medium for getting straws and grasses of the field essential for the livestock. None of sharecroppers continued labour relation with own previous *Haruwa-Charuwa* lords. Lalan told that previous landlords have also changed in terms of social relations and their behaviors. Most of old landlords have either died or migrated to other places. Some of their offspring have maintained good relations with *Mushahars*.

Damposiya Livestock (Livestock-Sharing)

When Lalan told the social life of *Damposiya* livestock in the *Musahar* community, it was observed that each household had at least one goat. At Jamdaha, most of the *Musahar* households had either *Damposiya* buffaloes, cows and goats or NGO- donated goats. Lalan told that peasants from southern village offered them goats and buffaloes on *Damposiya* system. The etymological meaning of *Damposiya* was derived from *Dam* (cash) and *Posiya* (feeding). The tenants actually feed grasses and grains to owners' livestock and during the conclusion of the agreement for sharing the raised livestock, their cash value is mutually decided. After feeding the livestock (i.e raising them), their market value will be increased. Lalan told that his grandchildren and children in the community attended irregularly in schools and some of them dropped out because of their role in raising the *Damposiya* livestock. Parents also preferred making their children *Charuwas* over schooling. I observed similar scenario when I was sitting at a school gate *Chaputra* with Ramkumar Sada. There was morning class at Saraswoti Higher Secondary School. Early morning, a group of students wearing blue sky shirts and blue paints/frocks was coming toward school gate and a caravan of livestock and *Musahar* school age children on the back of buffaloes were going toward opposite side for livestock-grazing. This was indeed an insightful observation. Ramkumar Sada told me," Look! children of *Tamang* and other communities are coming to school whereas children of the *Musahars* are following *Damposiya* livestock". He argued that *Damposiya Charuwas* contributed to family economy of the community. His argument boggled my old question: "Why didn't the *Musahar* children prefer to attend the school". Jhokilal answered, "Our community demands quick output rather than sustained outcome. School education does not contribute to everyday livelihood of the household". He even made fun out of unproductive school education. He added" We have problems in stomach and doctors give medicine of headache". For them, school education was neither directly connected with family economy of peasant household nor the children enjoyed school education.

In Nepal, domestic animals--usually cows, buffaloes and goats--can be leased in the same way as agricultural land, namely, for a share of the return. The owner of a young animal assigns it to a sharecropper to rear. When the fully grown animal is sold, the sharecropper receives a proportion of the sale price. In the case of a female animal, he also takes a share of the progeny. Goats and young buffaloes were reared on *Damposiya* system at Jamdaha. Jhokilal argued that goats were more beneficial in term of feeding and quick reproduction in comparison to buffaloes. Mother goats were accorded the status of "seed animals" and their progenies were divided equally between owner and tenant. He had been rearing two *Damposiya* goats. They reproduced four baby goats within a year. Two of them were taken by the goat owner and rest two solely owned by Jhokilal. He claimed that customer paid Rs. 10000.00 per each baby goat at weekly markets. Jhokilal also expressed dissatisfaction in relations with owner of goats. The owner always preferred male kids and if denied, he was interested to annul the agreement.

Both Lalan and Jhokilal agreed that the *Musahar* tenants of sharecropping relations were not classical tenant-peasants who did not bargain with the lords. Their relations were slightly influenced by aviability of cash loans in the time of crisis. Jhokilal said that it was easier to ask cash loans with the owner of livestock than with other people. He added that there was high interest among the *Madhesi* communities. Babur Sada borrowed Rs. 100,000.00 (one *lakha*) when his son migrated to Qatar in 2011. The interest rate of the loan was 60 percent per annum. After a year of labour migration, his son remitted Rs. 160,000.00 to pay back the loan. The loan was also not easily available for all *Musahars* at Jamdaha. The *Musahar* middleman was required for getting loans to layman *Musahars*. The middleman pockets some benefits from the transactions between the givers and takers. The middleman culture was also practiced in all kind of transactions like buying and selling of livestock, land, labour employment, getting loans and making important decisions in and out of the community.

Land on Lease

Karichan and Ayodhi Sada got involved in *Parwar* production (local name of *Trichosanthes dioica* –a kind of pointed gourd) in their *Jirayat land*. They started for their household consumption and last year they extended its production on 2 *Kattha*[4] land. They solved minor economic problems of the households by selling this vegetable. By observing the popularity of *Parwar* production in this area, few *Musahar* youths started vegetable farming on annual lease land. Jamdaha is a pocket area for *Parwar* vegetable production. According to Sijendra, *Parwar* cultivation is lucrative because it is regarded as prestigious and highly consumed vegetable in the *Madhesi* community. Sijendra was once *Haruwa*. He went to India for a year. He returned and started daily wage labour in the local community. He was not satisfied with the everyday wage labour. He had no cash to go to third country for employment. Then, he started *Parwar* cultivation on leased land. There was oral agreement between Santa Thapamagar, the landowner and Sijendra Sada. The latter had to pay Rs.4000. per year to the former. Santa was teacher. One of his brothers went to third country for employment and his father was a politician. Santa shared that he did not get wage labourer for the cultivation of land. Large area of land remained barren for more than five years and he planted mango saplings last year. Neither landowner nor labourers were interested on farming land. Sijendra took the land on lease because his wife and children could handle *Parwar* production when he was busy on contractual labour.

According to Sijendra, *Parwar* cultivation was not labour intensive. Once they planted *Padwar*, it continued production almost all seasons and lasted for three to four years. Bir Bahadur Lama, one of the largest *Parwar* producers of the area, shared that he earned more than Rs. 600,000.00

---

[4] *1 Kattha* = 0.0338 Hectare (and 20 Kattha = 1 Bigha).

from the *Parwar* production within a year. When I was sitting on school *Chaputra*, Bir Bahadur and Sijendra compared who got appropriate rate of that morning transaction of *Parwar*. Bir Bahadur knew that the day was festival (*Jitiya*) day among the *Madhesi* community. He sold Parwar on Rs. 35 per kg at *Golbazar* whereas Sijendra gave to a wholesaler on Rs.32 per kg near *Golbazar*. Their discussion concluded that market strategy was another important aspect to be learnt by the *Musahars* who had started to produce vegetable commodities for the market. Sijendra did not have much idea about commercial production. But he cultivated for household consumption and made small transaction at weekly market.

Contractual and Mediator Labour

In October 2012, I visited Raslal's home twice but I was unable to meet him. His wife told me that he was in jungle with wood cutters. I met him at my third attempt. We were familiar before this meeting. He was not only leader but also a labour contractor of Tetariya Musahari. He told that he stayed at wood cutting point for two nights. There was a contract between brick factory manager and Raslal to cut five trees. The oral agreement was on Rs. 5000.00 Then, Raslal informed youths with their leisure time about the wood cutting project. As a contractor, he pocketed Rs. 500.00 commission and rest of amount was equally divided among the labourers. Three labourers including himself finished the job within two days. He claimed that this type of contractual work was not only economically profitable but also a system of allowing the autonomy to the labourers regarding the pattern of the labour work (i.e the decision to engage the number of labourers and the duration of labour). Therefore, wood cutting project was their favorite choice among labour works available in this locality.

When Raslal was talking about contractual labour, an old man from *Yadhav* community came to make contractual agreement for paddy harvesting. The old man said," *Musahars* do not like to go for farm labour. Finding a God is easier than finding an agricultural labourer in *Madhes*. There are labourers in brick factories and construction fields but there is the dearth of labourers on farms. I will rather plant mango saplings next year." There was contractual agreement of paddy harvesting between Raslal and the old man. According to Raslal, each harvesting labourer will get 1 part out of 8 parts of his/her total harvested paddy. As a manager, Raslal will get either cash or kind commission from the land owner. The old man with murky face said that because of international labour migration and demand of off-farm labourers in the country, agricultural occupation was in crisis. Raslal argued that the *Musahars* preferred group work, work freedom and entertainment as integral part of labour regime. Their world view was collective living with minimum level of code and control. Similar type of view was also articulated in terms of landownership, settlement pattern, labour preference and migration.

When I was on *Chaputra*, Rarbin Tamang (son of intermediate landholder who has been living in Kathmandu since 2000A.D) came to find a labourer for preparation of *Parwar* plantation. His first question to the three *Musahar* youths was who they were. He told that he did not know half of young boys in the Musahari. The *Musahar* boys who returned from India (after their employments were over) were not familiar with Rabin because both parties mostly lived out of village. The growing unfamiliarity, which is characteristics of free market social relations (Roseberry, 1978), sets the setting of their bargaining. Rabin requested the *Musahar* youths who were sitting on the *Chaputra* for work. They pretended they were going for contractual labour. After many requests, one of them came to observe the land quantity and quality to dig. He responded that it would require four to five hours of work. Then, he demanded Rs. 500.00 but Rabin instead proposed Rs. 300.00 only. Both of them agreed on Rs. 400.00 for the work but the labourer did not begin the work. After an hour rest, the labourer returned back to the settlement. When he returned without saying any word, Rabin knew that the man was not willing to work. Similar story had been told by Yadhav friend at *Chorarwa* in 2012 A.D. I have observed similar type of denial of farm labour and preference of brick factory contractual labour at *Ghaletol* (Giri, 2012). Next day, Rabin went to Ramkumar Sada, the familiar labourer who became contractual agent of the settlement. Ramkumar shared with Rabin that young boys could not be handled by landowners.

The new generation was not ready to compromise for the continuation of traditional legacy of labour relations. Contractual labour market attracted the *Musahars*. They did not value agricultural labour. At evening, Ramkumar sent three womenfolk to dig the vegetable farm of Rabin.

Because of their extreme dislike of farm labour, most of them were interested to work on contractual work. Raslal was also labour contractor of two brick factories at Asanpur. Raslal Sada and Ramkumar Sada were leaders of contractual labour at Jamdaha *Musahar* community. Contractual labour is highly preferable work in the *Musahar* community. At Ghaletol, there were varieties of contractual works such as paddy harvesting, wood cutting, brick factory working, soil cutting and construction of houses and roads. The contractual labour was mediated by many middle contactors between buyers and sellers of labour. One of the interesting observations was that a farm owner consulted with a contactor for his cultivation and harvesting of crops. The contractor decided whole-sale price of the work. Then, the contractor visited labour settlement and lured the leader of the labourers by offering some alcohol and meat. The contractor also offered the works on contractual basis rather than day wage. Fixed amount of cash and bonus of alcohol and meat were offered. Then, the labourers tried to finish the given task much earlier. Time and days were not counted for wage. Similar contractual group works were also available in brick factories. The number of bricks was counted when they manufacture them or transport them. Wood cutting, house construction and paddy harvesting were also practiced on contractual basis. This form of labour was dominant in villages and roadside settlements of *Siraha*. Therefore, the *Musahar* labourers received work through a number of middlemen. Within the *Musahar* community, a few of them were labour contractors and many of them were mere workers. The case of *Musahar* is quite similar phase of transition of Latin American peasants in an age of nascent capitalism in 1970s. Peasants existed on land that was rarely adequate for survival, and so they diversified their economies through off-farm employments, contractual labour, commerce, and craft activities (Levine, 1977). As in the case of the *Musahars*, their peasant men were allowed to migrate for work, while the women attended to the small plot of land and to the forest that provide a safety-nets for the households.

Firewood as Commodity

After corn harvesting at Jamdaha, womenfolk from southern villages came to collect corn stalk for winter firewood/fuel. One morning I counted their number and found that more than 50 women and children carried corn stalk from *Musaharniya* corn field. It was a wastage for *Jamdaha-Musaharniya* people because *Chure* forest was close to them. The *Musahars* (especially women) and other marginalized communities collected firewood not only for their household use but also for commodity of weekly market at *Golbazar*. Because of firewood scarcity in southern villages, cow dung was wrapped around corn stalk and dried for cooking fuel. They did not have access to forest near their villages. There was high demand of firewood at weekly market. I observed that women from the *Musahar* community cut firewood in the *Chure* forest whenever they were free. They made bundles of firewood (of approximately 20 kg each). Gender dimension of firewood income was another area to be explored. Tulki Sada, one of the regular firewood collectors from the Jamdaha Musahari, told that there was not sufficient firewood in the *Chure* forest. By pointing the deserted land of *Chure,* she added that all trees grown nearby the settlements were cut many years ago. She told a number of wood cutting and firewood collection stories of the *Musahars* and non-*Musahar* communities. *Chure* forest was a visual bill-board to understand relations between state and the local community at Jamdaha. Many wood cutters were arrested and some of them were killed during the *Panchayat* and Maoist insurgency eras. After insurgency, government also did not pay much attention to control wood and firewood cutting in this *Chure* though special national project was implemented to preserve *Chure* under the presidency through "*Rastrapati Chure* Conversation Program". The local people expressed their revenge on the *Chure.* Though there were two community forestries at the bottom of the *Chure,* they were not well-managed. Out of the boundaries of the community forestries, local people would collect firewood freely. They said that government prohibited wood cutting in all *Chure* forest but they regularly collected and transported 40 to 50 bundles of firewood in each day. Tulki told that there was no firewood in the *Chure* and they had to walk at least an hour at another side of the *Chure* to find firewood. One person could hardly collect one to two bundles of firewood in a day.

At *Hatiya* (weekly market), the rate of a bundle of firewood ranged between Rs. 250.00 and Rs.300.00. I did not observe that any customer bought firewood on the basis of weight. They calculated the weight of firewood on the basis of observation and kind of wood tree. The rate was also fluctuated on the occasion of festivals, marriage season and quantity of consumption at the household level. There was a big firewood collection center at the roadside of Golbazar. The owner of collection center locally known as 'wholesaler' or middleman continued his buying and selling firewood business in all days. Therefore, firewood was significant source of income and alternative of livelihood among the *Musahars* of Jamdaha. Harilal Sada told that he had made home out of income of firewood sale.

Beyond Land, Peasant and Proletariat

In developing countries, the statuses of "peasant" and "proletarian", and the transition from the one type to the other have a clear contemporary relevance. The concepts of "peasant" and "proletariat" identify different forms of collective behavior, social organization and political action. The peasant is locally oriented and defensive politically, yet has contributed to revolutionary change. The proletarian looks for association beyond locality (given the fact that the process is the basis of modem class politics) but the political impact of the proletariat has varied with time and place (Goodman and Redcliff, 1981). In many parts of the world, the peasant disappears in the face of the modernization of agriculture or survives by combining agricultural work with non-agricultural labour migration. The transition from peasant to proletarian is neither final nor unequivocal. The proletarian also retreats in face of the decline of full-time wage employment in the cities, and the increasing importance of independent and part-time employment. Most critics of capitalism insist on the unity and homogeneity of the capitalist system and they argue that there is no longer a space outside of capitalism's empire (Hardt and Negri, 2000).

On the basis of agrarian livelihood, the *Musahars* were not peasants because they did not have private land. Tania Murray Li (2014) argued that indigenous Indonesian highlanders who were imagined by activists of global indigenous and peasant movements to be securely attached to their land and communities joined the ranks of people unable to sustain themselves. When highlanders joined the march of progress promised in modernization narratives, they lost traditional use and customary system of land relations which Li called Land's End (Li, 2014, p.2). For *Musahars*, land was common property resource. They denied owning land when landlords tried to register land in their names. Their stories indicated that they were interested to keep themselves away from state coercive power through regulations of land. When they were made *Haruwa-Charuwas*, narratives of agriculture experts of the *Musahars* were made. They followed the order of their landlords rather than practicing agriculture independently. They were never independent peasants in terms of their livelihood strategies.

The nature of work preferred these days by the *Musahars* could not only be limited into the box of "land", "peasant" and "proletariat". As global trend toward full-time agrarian subsistence and full-time wage labour appears to be ending, the *Musahars* were moving out of agrarian subsistence and toward deproletarianization (Portes et al., 1989). The shift of livelihood at Jamdaha was not from agrarian to industrial labour (full-time job) as in Europe. Those who followed peasantry were not peasants in classical sense. Therefore, their practice could be understood in Wolfean line of culture. In 'Europe and the People without History', Wolf argues, "A culture is . . . better seen as a series of processes that construct, reconstruct, and dismantle cultural materials, in response to identifiable determinants" (Wolf, 1982, p.277). But in contrary to classical Marxists, cultural reconstruction or transition from peasantry to (de) proletariat of the *Musahars* was not organized against landholding communities. The conclusion is that capitalist development has certainly transformed the *Musahars* of rural villages beyond recognition and differences, so that a general theory of peasant transition is not easily available at Jamdaha. Their traditional occupations and relations were at cross-road of new mode. They were free to choose varieties of livelihood and social relation options. As much as capitalism produces a singular grind of accumulation through dispossession, it is, as Tsing (2015) shows, a complex assemblage. In the rubble of that assemblage, salvage life continues to overflow and find its freedom, at least for a time (Tsing, 2015). But, their freedom is what Darren Byler (2017) called salvage freedom, in which freedom of salvage was not only restricted but also uncertain. The new course beyond peasant and proletarian way of *Musahars'* livelihood showed hopes of autonomy of work or at least freedom of choosing a particular form of livelihood activity.

Conclusion

Hitherto, the analyses of the "land" and "land relations" in Nepal have largely been influenced by the dominant narratives characterized by ethnocentric perspectives of policy makers and development activists. When the *Musahars* felt humiliation working on farm land, uninterested on agriculture and landed communities also changed land use pattern, land did not hold classical weightage of economic and political culture. The *Musahars* out of agrarian labour were found socio-economically sound and dignified. It does not mean that they do not need land but the told stories of land as prerequisite for the mainstreaming of the *Musahars* are questionable. The underlying interest and motivation of the *Musahars* were to get rid of farm labour. Their inclination toward contractual collective labour and migration articulated their sense of bewilderment without a sense of direction of the future course. This was the predicament of Lalan, Ayodhi, and Karichan, who could no longer sustain their families on the old terms but had no confident alternatives.

Classical anthropology's peasant as universal category was/is problematic because many communities like *Musahars* were/are not fit in any category: primitive cultivators, peasants and agricultural entrepreneurs. Anthropologists differentiated peasants and capitalist farmers on the basis of location and mode of production. It is observed that there are many types of peasant, varieties of proletariat and in between them, depending on the particular form of agricultural production and land tenure, and on the nature of industrialization. The *Musahars* did not have private land and agricultural inclination. Their understanding of land was different from landed communities' orientations. In some cases, they not only underestimated land-embedded power but also maintained certain level of alienation from private land and landed communities once there was homogeneous form of *Haruwa-Charuwa* labour regime which gradually changed into heterogeneous forms and contents in different places and different communities in east-central *Madhes*.

After economic liberalization and intense connectivity of global market, the plurality of proletarian peasantry increased not only differences but also deep inequalities among the *Musahar* labourers. Raslal and Ramkumar and many other youths remained contractual leaders who not only governed labour regime of the locality but also earned higher profits out of their roles in arranging the mass of labourers. The mass of *Musahar* labourers were in confusion when they were asked their occupation or future course of livelihood strategy. They did not have regular works and certainty of other works for coming days. This study refuted assumptions of classical Marxist analysis of capitalism in which rural peasants get transformed into proletariat. Everything, ruled by a singular capitalist logic of rational humans, is not found among the *Musahars* at pericapitalist space. With this interrogation on overarching capacity of capitalism, there might be space of study how capitalist and non-capitalist forms interact in periphery.

References

Appadurai, A. (1986). *The social life of things: Commodities in cultural perspective.* New York: Cambridge University Press.

Byler, D. (2017). Salvage freedom: Dialogues. *Cultural Anthropology*, website, June 8. *https://culanth.org/fieldsights/1137-salvage-freedom.*

Chaturvedi, V. (2007). *Peasant past: History and memory in the Western India.* Berkeley: University of California press.

Dhakal, S. (2007). Haruwa, the unfree agricultural laborer: A case study from eastern Tarai. *Contributions to Nepalese Studies*, vol. 34. (2).

Donaldson, W.J. (2000). *Sharecropping in the Yemen: A Study in Islamic theory, custom and pragmatism*. Leiden: Brill.

Giri, M. (2012). Marginalization and arts of resistance among the Musahars of east-central Tarai. In D.R. Dahal, L.P.Uprety and B.K. Acharya (Eds.) *Readings in anthropology and sociology of Nepal*. Proceedings of SASON Silver Jubilee Conference. SASON.

Giri, M. (2017). Political economy of kindhip and marriage among the *Musahars*. In L.P.Uprety, B.Pokharel and S. Dhakal (Eds.) *Kinship studies in Nepali anthropology*. Kirtipur: CDA, TU.

Goodman, D. and M. Redcliff (1981). *From peasant to proletarian: Capitalist development and agrarian transitions*. Oxford: Basil Blackwell.

Hardt, M. and A. Negri (2000). *Empire*. Cambridge, MA: Harvard University Press.

ILO. (1930). *ILO forced labour convention* (No. 29). ILO.

Jha, H. B. (1998). *Tarai Dalits: A case study of selected VDCs of Saptari District of Nepal*. Kathmandu: Action Aid Nepal.

Kahn, J.S. (1778). Marxist anthropology and peasant Economy: A study of social structures of underdevelopment. In J. Clammer (Ed.) *The New Economic Anthropology*. London: The Macmilan LTD.

KC, Bal Kumar, Subedi, G and Suwal, R.J. (2013). *Forced labour of adults and children in the agricultural sector of Nepal: Focusing on Haruwa-Charuwa in eastern Tarai and Haliya in far-western hills*. ILO Country Office for Nepal- series no. 11.

Kroeber, A. L. (1948). *Anthropology*. New York: Harcourt, Brace & Company.

Levine, D. C. (1977). *Family formation in an age of nascent capitalism*. New York: Academic.

Li, T.M. (2014). *Land's end: Capitalist relations on an indigenous frontier*. London: Duke University press.

NNDSWO. (2006). *Ethnographic study of Tarai Dalits in Nepal*. Lalitpur: Nepal National Depressed Social Welfare Organization.

Patterson, O. (1982). *Slavery and social death: A comparative study*. Cambridge: Harvard University Press.

Portes, A., Castells, M., and Benton, L. (Eds.) (1989). *The informal economy*. Baltimore: Johns Hopkins Univ. Press.

Prakash, G. (1990). *Bonded histories: Genealogies of labor servitude in colonial India*. New York: Cambridge University Press.

Roseberry, W. (1978). Peasants as proletarians. *Critique of Antrhopology*. vol. 11, Pp 3-18.

Sahlins, M. and Service, E.R.(Eds.) (1960). *Evolution and culture*. Ann Arbor: University of Michigan Press

Tsing, A. L. (2015). *The mushroom at the end of the world: On the possibility of life in capitalist ruins*. Princeton: Princeton University Press.

Wolf, E.R. (1966). *Peasants*. New Jersey: Prentice-Hall, INC.

Discourses on the Transformation of the Peasantry: Looking through the Life Histories from Dullav[1] Area

KAPIL BABU DAHAL

Introduction

Nepali society has been represented as 'agrarian' (Nepali & Pyakuryal, 2011) from the time Nepal was opened up for the outsiders and the planned development and social science writing in voluminous form began in Nepal after the democratic revolution of 2007 A.D. (Mishra, 1987). Various wings (e.g. MoAD, 2015) of the Government of Nepal (GoN) as well as social scientists working on and analyzing it and the international and transnational organizations (World Bank, 2002) construing the need of development and supporting Nepal's efforts to develop her are also portraying Nepal as an agricultural society and emphasizing on the centrality of agriculture in Nepali society. Development discourses have represented Nepal as a stable agrarian society (Seddon et al., 2002). Later on, they are joined by the numerous

---

[1] A collective name given to the study area which derives from a historical-mythical figure Dullav Lamichchane. There is an Undergraduate Campus and a Higher Secondary School in Ghyampeshal and their name also originate from him.

Non-Governmental Organizations (NGOs) which were established to further and complement the governments' efforts to develop the country. These NGOs were supposed to overcome the inherent weaknesses of both the public and the private sector (Pokharel, 2005). Nevertheless, these actors have construed the meaning of *Bikas*, the development in such a way that it is understood as something that "comes to local areas from elsewhere; it is not produced locally" (Pigg, 1992, p.499).

One of the remarkable commonalities of such developmental discourses is that most of them are inclined to portray Nepal as an agricultural, rural and traditional society (Nepali & Pyakuryal, 2011) consistently till now. This kind of over-simplistic generalization often connotes as if Nepali society, especially the rural one, is stagnant and has been static over the decades. Not only development agencies are inclined to construe Nepali society as perpetually static but also some scholars (e.g. Bista, 1991) who are critical in analyzing some aspects of Nepali society also join these folks to produce these types of stereotypical views. They do not talk explicitly on the changes taking place in agrarian relations, the backbone of economy with 33.1 percent of Gross Domestic Product (GDP) depends upon which and major source of food, income and employment for 65.70 percent of the population (MoAD, 2015). These orthodoxic representations ultimately overlook the changes occurring in the lives and situations of rural inhabitants. Departing from this kind of "exoticization" of Nepali agriculture and peasantry, based on empirical data from the central hills of Nepal, this paper clearly highlights the differentiation and emerging divergence in agricultural relation and thus, in the life of the rural inhabitants.

Migration has become a pervasive phenomenon in Nepali society affecting diverse categories of people. The historicity of migration (Mishra, 1987) in the central hills has been analyzed in light with its present situation, which will be crucial to look at alterations taking place in the migration and what kind of pattern so far has been emerged. Whether it is entirely the structural forces that shape the migration or one can see any agency elements in this process (Sharma, 2013) is also analyzed in this paper. Embedded with these issues, how migration is affecting the peasantry and peasants' social and cultural life is the prime focus of this paper.

Now, it becomes imperative to present the "real" picture of rural peasantry in Nepali society. With the help of micro level ethnographic data in light with the influential macro structural forces, this paper aims to portray how Nepali society is undergoing massive changes and how it has emerged during the last three decades. In other words, this paper sketches the changes taking place in the rural social milieu and people's way of living in the central hills of Nepal as observed, witnessed and experienced by the people themselves in their own life. What kinds of alterations are taking place in their vicinity and which ones are the "prime movers" for that? The massive alterations taking place in living conditions, livelihood options and land and agricultural relations make us think about whether it is a form of 'differentiation of peasantry' (Lenin, 1989). This concept developed elsewhere in their typical social and cultural context does not hold equal strength to capture such emerging realities of Nepali society.

This paper bases on ethnographic information generated through the collection of life histories for the study of 'Social Exclusion and Democratic Inclusion in Nepal', carried out in November 2008 in northern part of current Bhimsen Gaupalika and its vicinity in Gorkha district. During this phase of the study, a total of 15 life histories were collected from people of different age groups, who were from *Dalit*, *Gurung* and *Brahmin* caste groups. Later on, further information was added and updated over the years and mainly in the winter of 2016 about the pattern and changes taking place in life style, agricultural relations and migration. The informants were of the age group of 30 to 70 years old during the first fieldwork period and were inquired about the changes they had felt, observed and experienced in the last 20 to 30 years. Born in and being native of the area, I was frequently in touch with some of the research

---

[2] The study was carried out as a part of Social Inclusion Research Fund (SIRF) project in support of CMI, Bergen, Norway and it was conducted for CEDA, Tribhuvan University, Nepal. The research was conducted in Gorkha district in Nepal in November 2008.

participants, which helped to enrich the data with ethnographic details. I have experienced both the convenience and challenges because of my 'native' positionality. This helped me crucially to get access into the setting, and also to update data. On the other hand, 'making familiar unfamiliar' was a constant challenge in my attempt to generate information.

It is crucial to point out that the life history of a person has normally represented the time he has lived in, "so that his associates (in his native community could) identify him in most of the details of his life" (Simons, 1942, p.3 quoted in Blackman, 1992, p.1). By saying this, neither I am overlooking the possibility of typicality of individual cases nor underestimating the significance of subjective experience of an individual in his life time. I am simply emphasizing the significance of one's life experience to reflect upon and conjecture the nature and pattern of time he had lived in. As individual experiences and gives meaning to social phenomena as a member of the society, understanding of and meanings given to the life event represent the common pattern of the time that we are dealing with.

Understanding Peasantry from the Study Area

The word "peasant" came into in English in late medieval and early modern times. At that time, it was used to refer to the agricultural laborers, rural poor, rural residents, serfs, and the "common" or "simple" people. During that period, meaning of this term as a verb was to subjugate someone[3]. Bernstein (2003) states that the central analytical issue at stake is whether 'the peasantry' constitutes a general (and generic) social 'type' (entity, formation, class, and so on). That is, whether there are qualities of 'peasantness' applicable to, and illuminating, different parts of the world in different periods of their histories, not least the poorer countries of Latin America, Asia and Africa today, and their processes of development and underdevelopment.

---

[3] Oxford English Dictionary, "Peasant, N. and Adj.," *Oxford English Dictionary*, 2005, http://www.oed.com/view/Entry/139355?result=1&rskey=F232n4&.

For Bernstein, which many scholars have also found as a common thread to understand peasantry, the peasant family is both a production and a consumption unit. These peasants are small farm holders. Nevertheless, the peasant way of living and peasantry does not remain constant over the period of time. Scholars have argued that for various factors related within the peasantry itself and outside of, peasantry encompasses the element of uncertainty, which Harari states as:

> Concern about the future was rooted not only in seasonal cycles of production, but also in the fundamental uncertainty of agriculture. Since most villages lived by cultivating a very limited variety of domesticated plants and animals, they were at the mercy of droughts, floods and pestilence (Harari, 2014, p. 83).

Moderately steeply mountainous terrain form of land is found in the area. The area consists of most of the land as *Bari* (dry land) and some *Khet* (irrigated land). Two crops are basically grown in these lands in a year. Maize, millet and paddy are the major crops grown in the area. After the cultivation of maize and millet, in many clusters, *Bari* remains fallow for about four months. In few areas, people grow some vegetables like potato, onion, garlic, etc. as cash crops. Sometimes people also grow *Ghaiya Dhan* (rice grown in the dry land) in *Bari* by intercropping with maize. Summer paddy is cultivated only in *Khet*. Nevertheless, a new variety of paddy has been introduced in the area which could be grown in the dry land during rainy season. Local people stated that this variety is still at the experimental stage and will completely replace millet if it grows well. Buffaloes, cows and oxen are main forms of livestock. They are raised for milk and ploughing. Their dung is used as manure in the field. Now-a-days, the tendency of using chemical fertilizers has also increased. Goats, pigs, poultry and pigeons are raised for meat.

Dullav area comprised of mainly small landholding peasants. My research participants' narratives show that there was almost absence of absolutely landless families, the *Sukumbasis*, in the study area and also the landlords having huge quantity of land. It does not mean that Dullav area was an egalitarian society. Of course, there were some rich peasants and poor peasants. However, the land was distributed among many households. In earlier days, possession of the size of land and its productivity was the main criterion of determining affluence or its absence. Land is primarily transferred as per the patriarchal inheritance norm through the male lines.

Whether the scholars were able to see it or not, peasantry has not remained constant over the years. Drawing on from his discussions about the development of capitalism in Russia, Lenin (1989) stated that the total of all the economic contradictions among the peasantry constitutes what can be regarded as the differentiation of the peasantry. The peasants themselves very aptly and strikingly characterize this process with the term "depeasantising." This process shows the complete dissolution of the old, patriarchal peasantry and the creation of *new types* of rural inhabitants (p.173). Lenin acknowledged that the differentiation of peasantry is in the process of transition towards capitalism. Therefore, he accepted that the Russian community peasantry was not the antagonist of capitalism, but, on the contrary, were based on its deepest and most durable foundation (p.173).

In line with what Bernstein (2003) stated, some of the fundamental features of peasantry in the study area included the small landholdings among the peasants, use of traditional cultivation and harvesting technologies and thus inability of the subsistence farming of many families to feed the family members throughout the year. For their survival, up to some twenty five years ago, household head had to look for opportunity to buy food-grains from the peasant families having surplus produce. Buying cereals from others, *besaunu,* was a humiliating task in the agrarian society and people used to look upon the family who had to go for this option. Local shops used to sell only the commodities which were not produced in the area and no local shops used to sell food-grains.

At that time, it was very hard to find someone who can sell the cereals. There was no choice for a potential buyer to find different kinds of cereals. Actually, people used to look for millet, wheat and maize considering the relative cheaper price than that of rice and the unavailability of the latter. Even to explore the potential sellers, people used to visit some houses at night, when most of the children and women would have gone to beds or at least it was easier to hide from them in the dark. They had to find a generous one who would be ready to sell and not to disclose this reality to his family members and the others. Finding a willing seller used to become a matter of luck. Once the seller was found, then the household head used to bring the food-grains only in the dark at night or before dawn. Despite all his efforts, sometimes, the rumour of his *besaune* used to become talk of the locality by next morning!

Flourishing Migration

Some 20 to 30 years ago, many families who could not meet their subsistence from their family farms migrated out to different parts of *Tarai*, mainly in Chitawan. A few others did migrate to other parts of *Tarai*. Neighbors and relatives used to convince the emigrating families not to move away from the locality. However, they also could not offer viable alternatives to the migrating families to meet subsistence needs for staying in the locality. While moving out from the village, the migrating families usually left the village early in the morning, even before the dawn. The logic behind this was not to create difficulties of departure for both who are leaving and for those who are staying in the village. Nevertheless, these people, mainly the men, also visited the village time and again during the occasions of different festivals and rituals.

Bhag-Bhog Sakieko (time to use/consume is over) was the common phrase people used as a rationale for their movement to other parts of the country such as Kathmandu or Chitawan, the mostly preferred destination for the people from Dullav area. When there was a situation of *Bhag-Bhog Sakieko*, for whatever reasons, then the family could leave that place or give up property. Sometimes, this notion guided one's action and many other time this might have been used to justify the prior action (e.g. not to regret on leaving/selling a property). Most of the families which moved out to other parts of the country were the ones having larger number of family members.

There are arguments that population mobility is an increasingly important livelihood option for rural population (De Haan, 1997; Mosse et al., 2002; Seddon et. al, 2002; Bebbington, 1999; Whitehead, 2002). When conventional mode of agriculture was not able to meet the subsistence need of general populace in the study area, some people looked for alternatives to meet their survival needs. International labor migration for the local people also became a normal phenomenon as elsewhere in Nepal, which has a long history in Nepal and it has been a feature of livelihood strategies from the year 1851 which marked the Anglo-Nepal Treaty of Peace and Friendship that paved the way for 3000 Nepali soldiers to be inducted in the British Gorkha Regiment (DoFE, 2014). International labor migration, which began such a long time ago, has paved the space for Nepali society, as Breman (1996) states, in its transition to capitalism.

Nepal is not exception to many countries having a trend of people moving from rural to urban areas in search of livelihood opportunities and, within rural areas, to move out of agriculturally-based occupations, due to various reasons like growing pressure on natural resources, declining terms of agricultural trade and other broad ranging trends in society (UNDP, 2003; World Bank, 2002). An elderly Gurung's narration about his accounts of insufficiency of agriculture shows his attempts to expand livelihood options:

> I had not acquired enough land as my parental property. None of the occupations such as agricultural works, wage laboring locally, and long distance commodity portering for the merchants of Aarughat Bazaar could not overcome my miseries. To get rid of poverty, I moved to my wife's natal home but still the situation could not improve. I did not hesitate to acquire new skills regardless of my age.

People adopted seasonal labor migration, as Nepali and Pyakuryal (2011) analyzed in their study about the far-western Nepal, as the produce from the agriculture was not sufficient for the entire year. People used to go to different parts of India and few cities in Nepal when they find time after cultivation and harvesting of the farm produce. In such context, migration can be viewed as having 'structurally embedded agency' (Mosse et. al, 2002) to escape the constraints of agricultural limitations. Therefore, migration served as a livelihood strategy. The prominent feature of livelihoods approach is that it regards migrants as dynamic actors and not the vulnerable and helpless victims. They can devise enough tactics and cope with risks that may come across or imposed through external conditions (Whitehead, 2002).

Along with expansion of the road networks to district headquarters and closer to the villages, need of money increased gradually. When commodities from the expanding global capitalist market began to encroach in the area, the need of cash gradually increased, leading to the monetisation of economic activities. This increased drastically when a gravelled road was expanded up to the vicinity of Dullav area. Those who did not have off-farm cash income source found it difficult to manage their cash needs as their agricultural produce was hardly sufficient for their own need and not available for the sale in the market, even if there was any market. One middle aged *Brahmin* narrated his experience of how the sudden increasing need of cash affected him:

> I regret on not going for teaching or another source of cash income when it was easier to get job. I have experienced that money has taken the place of land. My economic situation has deteriorated these days, even though I have same quantity of land. Now-a-days nobody wants to invest in agriculture and people are fleeing from the village. Once interested to engage in expanding public works in the area, later on, I gradually gave up to make time for making money. For the expansion and improvement of my cash income source, I learnt some basic skills of carpentry and began goat trading. I have realized that making money and being educated are the two main aspects of the need of time.

De Haan (1997) studied Bihari migrants in India and which showed that income from migration for generations has provided an inseparable part of households based in rural areas. This makes it fairly apparent to us that migration is generally not a disjuncture in society's histories and is usually part of people's survival strategies. This applies also in the case of some *Gurungs* from the study area who worked in Indian and British army. Among these *Gurungs* joining the foreign military service was a lucrative profession not only to uplift their economic situation but also to raise their social status. This statement is bolstered by the following narratives of two male *Gurungs*:

> Grown up in an extended agrarian family, I began to support my family as a cattle herder. When I reached 16 years of age, I tried to get into Indian Army but could not become successful. My living condition improved only when my son got a job in Indian army. This made me able to give up wage laboring. Instead, I can hire people for our farm works these days (Gurung, aged 68).
>
> I served in Indian military service for 20 years and that helped me to lead a esteemed life back in my village. However, I am not happy that my sons could not become *Lahure* (Gurung, aged 72).

By picking up information from their memory, some of the local residents opined that one of the interesting things about the migration was that almost all the families which moved out were from the *Brahmin-Chhteri* and *Gurung*. Earlier, none of the *Dalits* moved out permanently at all. For them, the most likely reason could be that *Dalits* did not have adequate land to fetch some money to buy land in the destination area. Therefore, moving out permanently was not possible for the absolutely poor. Rather it was a choice for those who could afford to better their situation. Instead, many *Dalit* men went out for seasonal labor in India, mostly to work as factory laborers and security personnel. Only few of them were able to secure job in Indian military service. As it was not easier to get military job as a *Dalit*, some of them had changed their surname to other non-*Dalits*, which British and later on Indian government has recognized as "martial race" (Rand & Wagner, 2012) such as *Thapa*. They did not choose *Gurung* or *Magar* surname considering their physical appearance which resembles like *Brahmin-Chhetri*. Rather they opted for *Thapa*.

Seasonal labor migration during winter season in different towns in Nepal and in India (Delhi and Calcutta mostly) is not new phenomenon in the area. Men from all caste/ethnic groups migrated temporarily to find unskilled jobs. In the last 15 years, destinations of migrants have expanded immensely. Now-a-days, many people go to Gulf, South-East Asia and other few rich people go to Europe and USA. Up to 20 years ago (approx.), the remittance was heavily used as investment in buying both irrigated and non-irrigated land in their neighborhood. These days people do not invest locally. They prefer to invest in agricultural land and *Ghaderi* (land for constructing houses) outside of the village to move out permanently. Besides that, in contrast to few years back, now- a- days male migrants are taking their wives and kids along with them or in the towns. This has propelled the proportion of permanent migration of the youths and earners from the area. Thus, there has been the remarkable change in the destination of migration, number of migrants and use of remittance.

Until the dawn of the twenty first century, men emigrated to India to look for seasonal works. Very rarely any one went out of India except a few *Gurungs* who were able to join British army (DoFE, 2014). Those seasonal migrants returned back to their homes in various occasions and for different agricultural works including cultivation and harvesting. Most of the seasonal migrants used to leave their home after rice plantation with family consent. Usually, they used to leave in group of like-minded friends and stayed in the same city or locality even in India. If someone could not manage to come for crucial agricultural time or for some festivals, they would send cash and other sundry goods along with letters through other co-villagers or the neighbors. The cash and kinds brought by the migrants, the clean and new clothes worn by them and their descriptions of the mega cities often become the sources of inspiration for the rest of the youth folks in the village.

Macfarlane (1976) had described temporary migration pattern of the *Gurung* village, when many men from the village left for army service in British and Indian armies. These soldiers brought their pay and pensions back to the village and invested there. In the same *Gurung* village, after about 30 years, Macfarlane (2002) indicated that he saw significant changes in the nature of migration. In the recent years, he found that the retired servicemen settled elsewhere in the cities instead of returning to their village of origin. This kind of trend has rapidly increased in Dullav area as well. Many migrant's families had moved permanently mainly to Chitawan, Gorkha or Kathmandu.

Migration of youths and families from the locality has been propelled by the decade-long Maoist armed struggle as well. Many people left the village involuntarily to save their lives from both warring sides, the security forces and the Maoists. Families sold their lands and property at cheaper prices and bought some land with the money in Chitawan and other *Tarai* districts. Many youths left the village or did not come back to the village for a long time to escape the threats from either or both sides of the war and this new phenomenon created a situation in the village where one could find only elders, women and children. This trend continued even after peace agreement between the warring sides.

Changes in the pattern of migrants' communication with their family members had helped the former to keep in touch with the latter. Up to some 30 years ago, there was no alternative to letters and verbal messages from the side of migrants to their family members and vice versa. Many parents and families had to request a few educated people such as teachers to write and read letters on behalf of them to their sons or husbands. The scenario has changed a lot with increasing number of school going kids who can read letters for their mothers, grandparents and neighbors. With the mounting flow of less expensive imported Chinese tape recorders in the locality, migrants and their family members began to send their voices recorded in the cassettes to each other. This helped to minimize the dependence of migrants' family members to others in communicating with them. Over the last decade, mainly when the Maoist-led conflict ended with the peace agreement, the massive expansion of mobile phones in the locality has facilitated their family members to chat directly with them.

Changes in Agricultural Relations and Peasants' Way of Life

Agriculture, which is assumed to have begun some 12,000 years ago (Harari, 2014), had made a tremendous impact in the human society. The way agriculture is practiced and the social relations which agrarian activities encompass have also undergone significant changes. Dullav is not an exception to this phenomenon as it had also witnessed profound alterations in the agrarian relation in the last 30 years. It has been affected mainly by the mobility of local people and the developmental activities, mainly the kind of education imparted among the youths through formal education system. These factors have contributed to shifting agricultural relations to come out from the feudal mode.

Even before the establishment of two secondary schools in the locality, about 40 years ago, some *Brahmins* used to go to Banaras for their higher studies. Many schools were set up in the post-1990 period in different neighborhoods of the study area. Regardless of their literacy status, many parents began to send their kids to schools. After the completion of SLC, most of the graduates left the village for further studies in Kathmandu and other cities in Nepal. However, this modern education which had gradually propelled in early 1950s to meet the demands of the expanding government activities and the bureaucracies (Bista, 1991, pp.12-124), has not been able to meet need of the country even after more than six decades of its initiation. Still, there is a disjuncture between the education and need of the local people, especially in inculcating knowledge and skills to harness local resources and enhance the domestic production cycle of the country.

Modern education has contributed to produce youths and manpower disinterested to stay in their locality, sometimes even in their country, unable to mobilize the resources available in their surroundings and the graduates are inclined to look for jobs elsewhere in the governmental, private and developmental sectors. "Dichotomizing villages with that of *Bikas*" (Pigg, 1992, p.495), developmental efforts have also provoked people to leave undeveloped villages to look for better life. Linking job seeking mentality with the nature of education system, Pigg (1996, p.173) in her another writing had rightly stated that people in Nepal, at all economic levels, have a more immediate interest in the job prospects that the development activities may provide than in the actual improvements which these development programs are meant for. Moreover, she had pointed straightforwardly that such development and education propelled modernization has no link with the prevailing social reality:

Narrative of modernization …posits a rupture, a break that separates a state of modernity from a past that is characterized as traditional. Nepalis experience modernity through a development ideology that insists that they are *not* modern. Indeed, they have a very long way to go to get there (1996, p.163).

Pigg (1992) further adds that schools reinforce "the legitimacy of the opposition between *Bikas* and village" (p.502). Nevertheless, failure of our education system eventually contributes to the expanding and penetrating global capitalism as the former ultimately helps by supplying unskilled and semi-skilled laborers. As most of the youths are either actually emigrating out or see their future in international migration, imitating their fellow villagers, neighbors and relatives, this scenario gradually prepares them not to see their future in generating local production cycle (Mishra, 1987).

Propelled by a decade-long Maoist-led armed struggle, many youths, mainly men and families having surplus labor for agricultural works in the locality, left their locale for a long term or even permanently. Along with and related with this, many educated youths also did not return back to the village after completion of their studies in Kathmandu and other urban centers. These factors collectively contributed to the shortage of labor in the village. For instance, a 57 years' old *Brahmin* said, "Only elders and kids are there in the villages. There is a shortage of people even for funeral and wedding procession. People hardly find helping hands to bring patients to the hospitals as well".

One can see numerous fallow lands and abandoned houses in different settlements in the study area. Even within the vicinity of one of the earlier high schools in Tandrang, one can see more than a dozen of abandoned houses. The conventional beliefs of local people not to keep land fallow, property earned with hard efforts, was mainly based on the belief that abandoned *jaggale shrap dinchha* (land gives curse if it is kept fallow). Earlier, some people even asked for forgiveness from the god of the land (*Bhumi Deuta*) for not being able to cultivate the land, as 'people used to communicate with different components of the environment through Gods of different specializations (Harari, 2014). A villager from

Majhinthok reflects upon how keeping fallow land has become common practice in the area:

> Some 30 years ago, villagers claimed the marginal forest land through *khoria phadne* (slash- and- burn) to accumulate more land. These days, some families even keep their *gairi khet* (the well-irrigated land) also fallow. Many families gave up farming in the marginal land where they cannot have more produce. With the increasing forest around the settlements and declining farming in such marginal land, these days we have experienced *Bandar lagne* (i.e crop depredation by monkeys).

Up to a decade ago, out-migrant families used to find some other villagers as share-croppers, which is not possible these days as most of the youths have already left the village (regardless of their possession of their houses in the urban areas or not). Their elder parents rely on remittance for their living rather than producing food-grains from their own land. They have experienced that it is cheaper to buy rice and other cereals rather than cultivating them in their own land with the help of wage laborers. Wages rate which used to be Rs. 100/- for a man and Rs. 50/- for a woman in 2008 had increased to Rs. 600/- to Rs. 700/- for a man and Rs. 500/- plus for a woman in 2016. With such escalating agricultural wages, some families still continue their *parma*, a system of mutual exchange of labor, to cope up with this demand of money.

Boltvinik (2012) had rightly pointed out that agricultural capitalism can only exist when there are poor peasants, who are prepared to (and urged to) sell their labor seasonally (pp.14-15). This phenomenon had begun to emerge in my study area as well. Moreover, I would also argue that it is not only agricultural capitalism but industrial capitalism also needs peasantry with poor peasants for its sustenance. It is because of inability of the peasantry in Nepal that industrial capitalism in India, Gulf and South-East Asia get laborers, often cheaper, and for different period of times as per the need of the employer. Eventually, this kind of labor migration remains critical in pushing the society in transition to capitalism (Bremen, 1996).

Based on the analysis of secondary quantitative data, Regmi and Tisdell (2002) claim that migrants tend to send money back at home. They send 4.07 per cent of their income to their families. Even a small amount of remittance could contribute significantly to the livelihoods of poor people back at home. Blaikie et al. (2002) conducted a study to understand 20 years of social-economic change and continuity in central Nepal and claimed that the most prominent features of change was the increasing reliance on remittances sent back to rural households in the region from elsewhere.

To meet increasing cash need, besides migration, people also looked for different off-farm activities in their locality. The seasonality of agricultural works has barred many landless and small peasants to depend upon wage laboring in the locality. However, a few semi-skilled men have found carpentry and mason (especially working with cement) job as rewarding, having the wages of two to three folds more respectively. Increasing trend of youths to go out of the village has opened up new avenue of wage laboring for some people in the locality. A Gurung man in his mid forties narrates the need of diversification in the family income:

> I did portering and learnt carpentry for cash. I gave up working as a wage laborer only when my son became *Lahure* and advised me for that. In addition to his remittance, these days our family income has diversified and increased. Sale of local liquor, goats and milk in Ghyampeshal and pension of my mother-in-law from her *Lahure* husband are also sources of our family income.

There are many retail stores in Ghyampeshal and Arughat, which sell different consumer goods including agricultural products imported from other parts of the country and even from India and beyond. Not only in these two local towns, there are numerous retail stores also in various localities in Dullav area. Twenty years ago, these stores sold consumer goods which were not produced in the locality (for instance, cooking oil, kerosene, salt, some spices, cigarettes and tobacco, sugar, tea, biscuits and candies). They sold only few brands of noodles.

The increasing flow of remittance in the locality has brought alterations in various domains of life of these rural inhabitants. Reflecting the significant alterations in the consumption behaviors of local people, one can find recently added consumer goods in such stores in the locality. One can see various brands of sacks of rice in each retail store. Most of these rice sacks are from Nepal's *Tarai* region. Local retailers state that there is increasing demand of noodles and biscuits and some of the biscuits are made in India. Instead of a single brand of noodle previously available, reflecting the flourishing business of snacks in the overall country, there are different varieties of noodles produced in different parts of the country. The intensity and dimension of their linkage with the urban area has increased significantly and are reflected in their consumption behavior. For instance, a 53 years old Gurung remarked, "People buy cereals in the market and send their kids to private schools in urban centers. It has become a significant social marker of the family, aspirations of the parents and matter of family honor in the society". Similarly, sharing on the impact of transport system on the income-earning opportunities, a 58 years old Gurung said, "Newly expanded road – transport and the existence of local *bazaar* have increased our opportunities to raise income by selling milk and goats. Increasing income, construction of the road, education and expansion of the *bazaar* have changed situation of the locality".

Market has aggressively reached into different localities of the study area leading to alterations in the consumption habits of local people. People are giving up consuming bouquets and it has been replaced by rice. *Rice culture* has been expanded in the study area. These days, people have money to buy rice from the stores. When a son begins to make money, the first thing he asks his parents is " to give up consuming *dhindo* (traditional porridge) because now I can afford you to provide rice". With the escalating wages, cultivating cereals in one's own farm has also become expensive. When people have to buy, they go for rice. In this regard, a sexagenarian Gurung remarked, "Along with increment in the income, my family's dependence on agriculture has declined swiftly. Our food habit has also changed significantly. Now we do not eat millet *Dhindo* and consume only rice bought in Ghyampeshal. My son has strictly asked us not to eat *dhindo* anymore".

Changes in consumption habits have accompanied with the shifting values regarding buying cereal produce. This perfectly matches with the national level scenario of import of agricultural products of Rs. 40.15 billion in the fiscal year 2073/74 (Kafle, 2017). About 20 years ago, people used to ridicule a family if they buy cereals from others. Buying cereals from other households, which was not available in the local shops at that time, was termed as *Besaune*. Up to 20 to 25 years ago, people who used to consume cereals through this procedure were looked upon by their relatives and neighbors. *Besaune* was a fundamental indicator of the impoverishment in the conventional understanding. The family would go for *besaha*, cereals bought through *besaune* process, only when the family is on the verge of *anikal*, the famine.

Elder people are also aware of the increasing penetration of commercialization in their neighborhood. These days, people have to "buy everything including water as in the towns" (referring to the fee that they have to pay annually for the drinking water supply). Local people have experienced the trend of increasing consumption of meat. Up to 20 years ago, they consumed meat during *Dashain* and only in some other festivals such as *Saune Sakrati* (the first day of the month of Shrawan), *Maghe Sakrati* (the first day of the month of Magh) *Chaite Dashain* (Dashain festival which falls in the month of Chait). But now it has increased up to twice a week. Here, 'twice' might be metaphorical but it has to be taken symbolically.

These days, a few people have begun poultry farming in the locality. These farmers do not need to sell chicken elsewhere. Rather the buyers come into their farms. Along with the increased availability of cash in the locality, people have begun to consume more commodities bought from the market. Moreover, the value system has been inclining gradually towards the consumerism. Thirty-nine years old Mrs. Khanal (whose husband works in South Korea over the last two years) stated this in her own words: *"Afano aaatma marera ko kamauncha? Sampatti bhaneko ta khaera laera banki raheko po ho ta"* (Who will opt to earn at the cost of one's inner feeling? Wealth is what is left after spending on your food and clothing). Her narration also signifies the primacy of food and clothing over other commodities. This newly inculcated value system juxtaposes with that of the age old values that used to emphasize on *puryaune/tarne* among the peasants in the study are. *Puryaune/tarne* signified the primacy of minimizing expenditure and managing with what you have at hand. People consider that increasing availability of money in the locality is the prime cause behind such emerging shifting trend.

Some 25 years ago, there was no practice of selling fruits and vegetables in the locality. Any surplus would have been shared among the relatives and neighbors. Usually, relatively better- off family used to offer such surplus things to others as they could afford to utilize land for that as well. Those who would get such fruits and vegetables would have the feeling of "generalized reciprocity" (Wentowski, 1981) who would repay, over the time, in different forms. Such repayment could be another variety of fruits and vegetables, usually less than a day long free labor contribution for some works, or providing some gifts such as chocolates and biscuits brought from *Lahur* to the family and so on.

Older people express their nostalgia to those good days remembering cooperation and selflessness among the villagers and express their dissatisfaction with the expanding selfishness. Paradoxically, they also feel sorry about the suspicion of those days when people had to have legal documents even for small amount of money (like even for Rs. 10.00). They think, earlier, money lenders were merciless who used to impose exorbitant interest rate in addition to some other non-cash obligations of the borrowers (e.g. demanding free labor contribution, he- goats etc.)

The contesting values towards the land has been a historical phenomenon in the locality as it enjoyed both the sacred status (*Dhartimata*- the 'mother land' as source of food and survival) as well as a status of commodity to be sold in the market. The pace of commodification is becoming omnipresent leaving far behind aesthetic value of land. However, during the last 20 years, transaction in land has declined significantly (be it *Khet* or *Bari*). Initially, this triggered to increase the tendency of share-cropping and later on, the practice of leaving fallow-land was resultant. Increment in agricultural wages, migration of the youths, and increasing remittance has collectively contributed for that.

Along with the alteration in agricultural relations, one can see the symptoms of declining significance of land in some avenues of their social life. There was a practice of *Tauwa herne* (i.e to observe the storage of dry straw to decide whether to arrange a daughter's marriage to the family or not). Besides *Tauwa*, bride's family also paid attention to the distance to the drinking water-tap from the household. Obviously, the longer distance negatively affected the possibility of arranging the marriage. It was because the work burden to the bride who would be married out into such locality would eventually increase because of longer time and distance to fetch water. Now the place of *Tauwa*, an indicator of prosperity, is taken of by the "educational attainment", "earning capacity" and "the quality and status of groom's employment".

The way people understand affluence locally has been changed significantly over the years in the locality. A septuagenarian Gurung man from Ghyampeshal area nicely reflected upon what he has experienced in his life time about the changing notion of how people understand the middle class in the context of declining value of agriculture. He remarked:

> The criterion of richness has changed these days. Earlier, it was the "self-sufficiency " or "surplus" in agricultural products which has now been replaced by the "size of income" of the family and its "state of financial resources" to afford to buy rice and other consumer items. My family was of middle class when I was 15 years old because we did not have to buy food-grains. Now it is also middle class because we can afford to buy consumer items from the market.

Conclusions

Capitalism and its expanded wings (such as consumer markets, global labor markets and international labor migration and spreading out of road networks devoid of tar-macadam in the rural hinterland) are creating spiral effects on the rural people's living conditions and their ways of life. In fact, international labor migration has become a triggering factor for social change in rural parts of Nepal and especially to accelerate their transition into capitalist mode of living. There are long-run consequences on the migrants' families and their value systems, ultimately leading to the alterations in their ideas and practices related with consumption, relation and attachment with the land and social relation in general within the society.

In fact, at least in the beginning, labor migration has been propelled by the inability of agriculture in meeting subsistence needs of the people. This is one of the reasons for the gradual loss of importance of agriculture among the youths, who were more educated than the people of their parental generations, and thus their attraction was towards migration. With the increasing need of cash in monetized economy, which has become a basic requirement of global capitalism for the transfer of resources and profit, youths from the locality gradually began to see their future in migration. Unlike their predecessors who used to go to India, presently youths chose to go to Gulf, East Asia and beyond. The function of the failure of Nepali education system to be able to see the viability of goodness in one's own surroundings eventually ends up with the functional value of contributing to the expansion and perpetuation of unequal global capitalist system, maintaining to keep Nepal in disadvantageous situation, through the production of cheap labor for the economic growth of the other developed or developing countries. Our education policies adopted in the 1950s seems to be in harmony with the need of global capitalism as it had also emphasized that the contribution of education should be able not to remain in isolation in the world which has come to us (MoE, 1956, p.2). Maoist-led decade-long war had also added to further the process of capitalist expansion into the study area as it also contributed in accelerating migration, including international labor migration.

International labor migration has supported global capitalism to acquire human resources with different levels of skills for the job unwanted for the inhabitants of the developed countries. And the remuneration that these immigrants send home as remittance further contributes to expand its market base for the commodity which eventually paves the way to extract monetary resources from the global south. Beyond the predictions of some world systems theorists (e.g. Wallerstein, 1974) and dependency theorists (Frank, 1967), the global capitalist centres are also expanding even in our surroundings (China and India).

International labor migration has expanded people's imaginations even in their everyday matters and also for crucial life events. Their worldview regarding their livelihood options are not confined to their locale only; rather they can think of in terms of the global context, whether to look for job at the local level or go to India, Gulf, South-East Asia, Europe or USA. Likewise, people do not confine themselves to consuming goods produced locally, but can choose from those which are produced in other countries and available in the local markets. These two crucial aspects of their life, and livelihood options and consumption make them different from their own immediate predecessor peasants (not only their parents but also from themselves of some 30 years ago). Even an old couple can be found somehow involved in planning international migration of their sons and grandsons (though very rarely for the daughter and grand-daughter). This kind of expanded imagination and possibilities makes the situation of Nepali rural inhabitants distinct from what Lenin (1989) had analyzed as how the differentiation took place in the Russian peasantry.

The way peasantry is undergoing changes in Nepali society differs from what Lenin had assumed while analyzing Russian peasantry-- through the loss of land to the emerging capitalistic landowners or the others had grabbed the land, whereas in Nepali case, it appears more in the mid-path of being-in-the-world (Heidegger, 1962, p.52), through people's "choice" of migration which shaped their preparation to transient to the capitalist societal structure. Moreover, unlike what Lenin was unable to see or what he had underestimated the role of migration, especially international migration, and inappropriate (rather than incapable and inefficient) education as it was blind to mobilization of local resources. As most of the narratives from the life history show parents view that their kids will have better job (not income sources) because of their education. Ironically, none of them referred or even aspired that their kids will involve in mobilizing local resources and creating jobs for others. Rather, the successful one with "good education" will be a good job seeker, implying that he will go outside, look for job from others and may not be confined to the village. Likewise, unlike what Lenin has envisioned as the emergence of entrepreneur as rural bourgeoisie and agrarian laborers as rural proletariat from the disintegration of conventional patriarchal peasantry, peasantry in Dullav area is disintegrating, gradually losing its attachment to the land and gradually migrating out to the neighboring towns and cities. To conclude with, changes taking place in Nepali peasantry does not resemble what Lenin has assumed capitalist changes will take place in peasantry. While looking at the people's aspirations and present objective agrarian reality, the emerging scenario does not look like differentiation of peasantry, rather it grossly looks like unattractiveness of agriculture—a sort of new transformation in the traditional agrarian economy. As above analysis reflects, subjective factors related with the individuals and households and structural forces shaping these subjective realms are contributing to creating the situation of repulsion from agriculture. Indeed, the conventional "doxa" (mindset) of the members of Nepali peasant society that "educated youths should look jobs outside the ambit of agriculture" and failure of the state's agricultural polices for the "modernization of agriculture" to pave the path for the "agro-industrialization" (despite the formulation of a plethora of donor-driven polices for the commercialization of agriculture) have been the triggers failing to attract the educated youths to be the "agro-entrepreneurs" within the country. Nonetheless, of late, a new propensity among the educated youths with overseas exposure for adopting "modern agriculture" is laudable because they have the motive to continue it as a "dignified profession" (i.e educated ones can opt agriculture as a lucrative profession with dignity) by producing for the makets (both internal and external). This inculation of professional values among the larger segment of the youths may trigger the attraction towards the "market-oriented agriculture", paving the path for the further transformation of Nepali peasant economy.

References

Bebbington, A. (1999): Capitals and capabilities: A framework for analyzing peasant viability, rural livelihoods and poverty. *World development,* 27(12), 2021-2044.

Bernstein, H. (2003). Farewells to the peasantry. *Transformation: Critical perspectives on Southern Africa,* 52 (1), 1-19.

Bista, D. B. (1991). *Fatalism and development: Nepal's struggle for modernization.* Calcutta, India: Orient Longman.

Blackman, M. B. (1992). Introduction: The afterlife of the life history. *Journal of Narrative and Life History,* 2(1), 1-9.

Blaikie, P., Cameron, J., & Seddon, D. (2002). Understanding 20 years of change in West-Central Nepal: Continuity and change in lives and ideas. *World development,* 30 (7), 1255-1270.

Boltvinik, J. (2012). Poverty and persistence of the peasantry. Background paper for international workshop on poverty and persistence of the peasantry. El Colegio de México (Pp. 13-15).

Breman, J. (1996). *Footloose labor: Working in India's informal economy.* London, UK: Cambridge University Press.

De Haan, A. (1997). Unsettled settlers: Migrant workers and industrial capitalism in Calcutta. *Modern Asian Studies,* 31 (4), 919-949.

DoFE. (2014): *Labor migration for employment. A status report for Nepal: 2013/2014.* Department of Foreign Employment (DoFE), Kathmandu.

Frank, A.G. (1967). *Capitalism and underdevelopment in Latin America.* London, UK: Monthly Review Press.

Heidegger, M. (1962). *Being and time* (trans. J. Macquarrie & E. Robinson). New York, NY: Harper & Row.

Harari Y. N. (2014). *Sapiens: A brief history of humankind.* Toronto, Canada: Penguin Random House.

Kafle, R. (2017). *Aath bastuko aayatma paune 6 kharba kharcha* (Quarter to six billions spend on 8 imported goods). Aarthik Abhiyan national daily July 30, 2017.

Lenin, V. (Ed.). 1989. The differentiation of the peasantry. In *Development of capitalism in Russia* (Pp. 172-186).

Macfarlane, A. (2002). Sliding down hill: Some reflections on thirty years of change in a Himalayan village. *European Bulletin of Himalayan Research.* 20, 105-124.

_____. (1976): *Resources and population: A study of Gurungs of Nepal*. London, UK: Cambridge University Press.

Mishra, C. (1987). Development and underdevelopment: A preliminary sociological perspective. *Occasional Papers in Sociology and Anthropology*, 1, 105-135.

MoE (Ministry of Education). (1956). *The five-year plan for education in Nepal*. Kathmandu: Bureau of Publications, College of Education.

MoAD (Ministry of Agriculture and Development). (2015). *Nepal portfolio performance review*. Kathmandu, Nepal: MoAD.

Mosse, D., Gupta, S., Mehta, M., Shah, V., Rees, J. F., & Team, K. P. (2002). Brokered livelihoods: Debt, labor migration and development in tribal western India. *Journal of Development Studies*, 38(5), 59-88.

Nepali, P. B., & Pyakuryal, K. N. (2011). Livelihood options for landless and marginalised communities in an agrarian society: A case study from far western Nepal'. *Pakistan Journal of Agricultural Science*, 48(1), 1-10.

Pigg, S. L. (1992). Inventing social categories through place: Social representations and development in Nepal. *Comparative Studies in Society and History*, 34 (3), 491-513.

_____. (1996). The credible and the credulous: The question of "villagers' beliefs" in Nepal. *Cultural Anthropology*, 11(2), 160-201.

Pokharel, T. (2005). Developmental practices in Nepal: A proposal and a critique. *Perspectives on Society and Culture*, 1(1), 35-38.

Rand, G., & Wagner, K. A. (2012). Recruiting the 'martial races': Identities and military service in colonial India. *Patterns of Prejudice*, 46(3-4), 232-254.

Regmi, G., & Tisdell, C. (2002). Remitting behaviour of Nepalese rural-to-urban migrants: Implications for theory and policy. *Journal of Development Studies*, 38(3), 76-94.

Seddon, D., Gurung, G. & Adhikari, J. (2002): Foreign labor migration and remittance economy of Nepal. *Critical Asian Studies*, 34 (1), 19-40.

Sharma, J. R. (2013). Marginal but modern: Young Nepali labor migrants in India. *Young*, 21(4), 347-362.

UNDP (United Nations Development Program). (2003): *Human Development in South Asia*. Wallerstein, I. (1974). *The modern world system: Capitalist agriculture and the origins of the European world economy in the sixteenth Century*. New York, NY: Academic Press.

Wentowski, G. J. (1981). Reciprocity and the coping strategies of older people: Cultural dimensions of network building. *The Gerontologist*, 21(6), 600-609.

Whitehead, A. (2002). Tracking livelihood change: Theoretical, methodological and empirical perspectives from north-east Ghana. *Journal of Southern African Studies*, 28(3), 575-598.

World Bank (2002): *Rural development indicators handbook for the world development indicators*. Working Paper, Washington DC.

Peasants' Land Rights: A Perspective from Engaged Anthropology

LAYA PRASAD UPRETY

Perspective from Engaged Anthropology

One of the most memorable lines of Karl Marx is his assertion that "The point is not merely to understand the world but to change it". With regard to anthropology, the role might be re-phrased through the formulation of three naïve questions which may be read as follows, "Is the role of anthropologist to try to change the world or to merely understand it?; Can (and should) anthropologists work as advocates for the rights of people they study or does this compromise their objectivity?", and "should anthropologists engage as active agents of change?". These questions must be critically rethought in the world which is characterized by the unequal power relationships (Kelletm, 2009, p.22).

Forms of engagement of anthropology extends from basic commitment to our informants to sharing and support with the communities with which we work, teaching and public education, to social critique in academic and public forums, to more commonly understood forms of engagement as collaboration, advocacy and activism. The reason we are interested in engaged anthropology is that we are committed to an anthropological practice that respects the dignity and rights of all humans and has a beneficial effect on the promotion of social justice. The contribution of Nancy Scheper-Hughes is relevant here who proposes "militant anthropology" because for her, cultural relativism is no longer appropriate to the world in which we live and that anthropology, if it is to be worth anything at all, must be ethically grounded. With this moral claim of hers, anthropological writing can be a "site of resistance". There is the need now to develop a politically-engaged anthropology. Anthropologists can offer the "social critique" which, in its broadest sense, refers to the anthropological work that uses its methods and theories to uncover power relations and the structures of inequality by examining the structural factors temporally. Anthropologists can produce "collaborative ethnography" which is the powerful way to engage the public by making to assist local communities in organizing efforts, giving testimony, acting as an expert in courts, witnessing human rights violations, serving as translators between community and government. They can also conduct the "activist research" by affirming a political alignment with an organized group of citizens in struggle and allowing dialogue with them for determining each phase of the process... The implications of ethnographic detachment in a world characterized by "resource inequalities", "land grabs", and "political violence" are problematic. Under these circumstances, "ethnography cannot be apolitical". There is the room for anthropologists to be engaged in activism/advocacy by drawing on persons' knowledge and commitments as citizens or as humans confronting the violations or suffering of other humans. The concept of the "anthropological citizen" proposed by Barbara Rose Johnston is important which connotes that the anthropologist can serve as a scribe, documenting abuses and as adviser, advocate, and partner in advocacy. Contemporary anthropological activism takes the position that anthropologists are "responsible for the potential effects of the knowledge produced about people and their cultures, to contribute to decolonizing the relationships between researcher and research subjects, and to engage in a form of anthropology that was committed to human liberation (Low and Merry, 2010,pp. 203-226). Anthropologists have to work for "social justice" so that all the rights of human (be they civic, political, economic, social, and cultural) are realized to the fullest possible extent in the world.

In June 1999, the American Anthropological Association (AAA) adopted the "statement of human rights". Its preamble reads, "Anthropology as a profession is committed to the promotion and protection of the rights of people and peoples everywhere to the full realization of their humanity, which is to say their capacity to culture (given the fact that the capacity of culture is tantamount to the capacity for humanity)…Anthropologists should continue to be concerned whenever difference is made the basis for a denial of basic human rights, where "human" is understood in its full range of cultural, social, linguistic, psychological, and biological senses" (humanrights-americananthro-org/1999-statement –on-human-rights). In a way, the basic contents to human rights seem to be consistent with international principles. It is, therefore, important for anthropologists to be engaged in the debate on enlarging our understanding of human rights on the basis of anthropological knowledge and research. Definitively, "advocacy anthropology" " or "engaged anthropology" can be considered as a sub-field within applied anthropology because anthropologists have the tremendous role to play as "social critic" (Susser, 2010) to interrogate the structural inequalities of the policy and practice by following a critical trajectory for anthropological engagement (Gonzalez, 2010) in the production of critically engaged ethnographic practices (Clarke, 2010). Granted this perspective of engaged anthropology, "land rights" of the poor peasants can be analyzed as the "human rights" because they are to be mainstreamed in the social world which is historically and structurally characterized by the "resource inequities".

The Purpose of the Paper

The main purpose of this approach paper is to provide a framework for linking land as human rights to help facilitate the engagement process of civil society organizations (CSOs) with regional networks and national human rights institutions (NHRIs) in South Asia. It also outlines the potential areas of collaboration between NHRIs and CSOs in promoting land rights and protecting land defenders at national and regional levels.

Land Rights and Human Rights: A Connecting Framework

This section explores some concepts and theoretical considerations on human rights, land rights and relationship of these two main concepts, main human rights standard mechanisms and international conventions/agreements and the place of land rights on them.

Land Rights in the Light of Human Rights Concepts

Human rights are the rights that humans have and are entitled to simply by virtue of being humans. They are inherent and inalienable rights that human beings require to live a dignified life. Collectively, they are comprehensive and holistic statements (PWESCR, 2015) elaborated and codified in the United Nations (UN) Declaration of Human Rights which had the explicit exposition of the civil and political rights of human beings (as adopted and proclaimed by the General Assembly Resolution 217 A-111- on 10 December 1948) and the International Covenant on Economic, Social and Cultural Rights (ICESCR) ratified and acceded by the General Assembly Resolution 2200A (XXI) on 16 December 1966 which has been in enforced since 3 January 1976. Succinctly put, all human rights – civil, political, economic, social and cultural – are recognized as universal, inherent, inalienable, indivisible and interdependent body of rights (PWESCR, 2015).

The preamble of the resolution of the UN Declaration of Human Rights clearly states, "The recognition of dignity and of equal and inalienable rights of all members of the human family is the foundation of freedom, justice and peace in the world". It further adds, "People of the UN have in charter reaffirmed their faith in fundamental human rights, in the dignity and worth of human person and in the equal rights of men and women and have determined to promote social progress and better standards of life in larger freedom". Together with the enjoyment of freedom of speech and belief and freedom from fear (proclaimed to be the highest aspiration of the people), there has been the pledge of the member states to achieve the promotion of universal respect for and observance of human rights and fundamental freedoms. Of the 30 Articles pertaining to civil and political rights, Article 17 is exclusive on right to property (under which people must not be arbitrarily deprived of it). Article 25 is on right to a standard of living (under which the well-being of the family includes food) (UN General Assembly, 1948).

Succinctly put, economic, social and cultural (ESC) rights mainly include the right to self-determination, equality, non-discrimination, gainful work, just conditions at work, social security, health, education, food, water and sanitation, housing and cultural rights – all essential for one to live a life both with dignity and freedom. The essence of the preamble of ICESCR (1966) also emphasizes the recognition of the inherent dignity and of the equal and inalienable rights of all where the human family is the foundation of freedom, justice and peace. More importantly, it also recognizes that the ideal of free human beings enjoying freedom and want can only be achieved if conditions are created whereby everyone may enjoy his/her economic, social and cultural rights, as well as his/her civil and political rights (UN General Assembly,1966).

More specifically, as indicated above, all people under ICESCR have the right to self-determination which allows them to make their own decisions politically for the free pursuit of their economic, social and cultural development. People have the freedom to dispose their wealth and resources. They cannot be deprived of their means of subsistence under no circumstance and the state parties are required to promote the realization of the right of self-determination. Emphasis has also been placed on the progressive realization of the rights by the adoption of legislative measures (for the enjoyment of rights without discrimination). Effort to undertake to ensure the equal right of men and women to the enjoyment of all economic, social and cultural rights has been prioritized. Article 11 of ICESCR writes that the state parties are required to recognize the right of everyone to an adequate standard of living for himself/herself and his/her family, including adequate food, clothing, housing and to the continuous improvement of living conditions to ensure the realization of this result. It also recognizes the fundamental right of everyone to be free from hunger for which the state parties, at times individually, and at times through international cooperation, must adopt the measures including reforming agrarian systems in such a way as to achieve the most efficient development and utilization of natural resources. Article 15 recognizes the right of everyone to enjoy the cultural life/rights (UN General Assembly, 1966).

Against the above backdrop of the indivisibility of human rights, we can argue for 'land rights as access to food', and 'land rights as rights to housing' because legal security is seen by ICESCR as a key element of the right to 'adequate housing'. In this regard, two aspects are to be considered: (i) positive aspect under which land rights are considered to be an essential element for the achievement of the right to adequate housing, and (ii) negative aspect under which land dispossession would qualify as forced eviction in direct violation of the right to housing. Indeed, the focus on security of tenure and access to land as essential elements of the right to adequate housing is also a central feature of the UN Special Rapporteur on adequate housing. Indeed, land right can be considered an 'accessory right' to the realization of other human rights (Jeremie, n.d).

Discussing land rights issues in international human rights law, Elisabeth Wickeri and Anil Kalhan (2010) have argued that up to one quarter of the world's population is estimated to be landless, including 200 million people living in rural areas. For many of these people, the condition of landlessness threatens the enjoyment of a number of fundamental human rights. Access to land is important for development and poverty reduction, but also often necessary for access to numerous economic, social and cultural rights, and as a gateway for many civil and political rights. However, there is no right to land codified in international human rights law. Land is a cross-cutting issue, and is not simply a resource for one human right in the international legal framework. Rights have been established in the international legal framework that explicitly relates to land access for particular groups such as indigenous people and, to a more limited extent, women. In addition, numerous rights are affected by access to land, including the rights to housing, food, water and work, and general principles in international law also provide protections relating to access to land such as equality and non-discrimination in ownership and inheritance (Wickeri, and Kalhan, 2010).

Effects of illegal land-grabbing are the violations of human rights (social, cultural, and economic rights), forced evictions, land disputes and deprivation of food, shelter and water for the land-poor people. Land concessions, land disputes and evictions threaten the existence of indigenous peoples who have used the lands for centuries because they do not use the system of individual property, but rather own the land collectively (Virak, 2014). Land-grabbing is a source of poverty, a source of urbanization and the source of malnutrition (Charya, 2014). In Gandhian perspective, people do not own the land, but they need to manage its sustainability to govern itself. This shows the rationale to look at land rights as human rights even for the community mission to sustain land for generations. Land rights can be regarded as human rights. The source of human rights, equality and land right is mainly from the responsibility to sustain lands with proper planning, management and control (Bhardwaj, 2014).

Land rights can be seen from the human rights perspective by analyzing their relationships. Various legal frameworks and international agreements and conventions have mentioned the importance of protection, maintenance, and respect to people's land rights in order to achieve sustainability and prosperity of the people, both at local and global levels. Experts, including the Special Rapporteurs of the UN Commission on Human Rights, pointed out that, "land rights is a gate to maintain certain human rights such as the right to water, the right to adequate housing, the right to health, the right to adequate standard of living, the right to food, and other rights"… Any violation, grab and eviction that cause loss of people's right over land creates an agrarian conflict which is also related to certain human rights violations. From the human rights perspective, there is indivisibility of human rights and, therefore, the state has the obligation to guarantee the security of land tenure in order to promote and protect the economic, social and cultural rights. They, in turn, contribute to protect civil and political rights (Bachriadi, 2014). The Phnom Penh Communiqué on Land Rights and Human Rights in ASEAN at the conclusion of the workshop on the theme 'Mainstreaming Land Rights as Human Rights" (16-17 September 2014) contained the following main aspects of emphasis on: (i) respecting and upholding human rights, including indigenous rights; (ii) addressing structural injustices at regional and national levels (vis-à-vis land rights), and (iii) ensuring access to justice, appropriate remedy and redress of the people's grievances (vis-à-vis land rights).

The Asian People's Land Rights Tribunal Forum was held on 16-17 January 2014, Malcom Hall Theater, University of the Philippines, Diliman, Quezon City, Philippines. It was a venue for land grab victims to demand accountability of respectable institutions as well as to raise the public awareness on the violation of smallholder rights within land investments happening in the Asian context. The tribunal has viewed that land grab cases merit a re-examination from the perspective of human rights because charges of human rights violations cannot be ignored, particularly involving corporations and other business enterprises in which powerful local and foreign interests are involved. Its Diliman Declaration has called upon states, institutions, and corporations concerned to respect and adhere to the following major principles: (i) respect and uphold human rights and environmental standards and commitments (by endorsing international declarations, ratifying multilateral conventions, developing policies, addressing human rights violations, respecting, upholding indigenous peoples' (IPs) rights, regardless of whether they are recognized by the state, including their customary laws and rights, traditional institutions and decision-making processes, and community protocols and procedures, engaging with IPs, and local communities as rights-holders, protecting human rights and recognizing forced evictions as gross human rights violations; (ii) address structural injustices at the national and sub-national levels, and (iii) ensure access to justice, appropriate remedy and redress of people's grievance (Diliman Declaration, 2014).

The above analysis by the scholars clearly shows that violation of land right of people leads to the violation of human rights (be they civil and political or social, economic and cultural). Indeed, land right is the accessory right for the actual realization of other human rights. A case in point is access to government services in Nepal – water, electricity, banking services, citizenship certificates which are vital for the enjoyment of voting rights and also serve as the gate to acquiring passports for gainful foreign employment which constitutes 25 percent of gross national income.

Critique: Indeed, both the UN Declaration on Human Rights and ICESCR have failed to address the issues of peasants in general. If critically viewed, the UN Declaration on Human rights touched on the issue of property, thereby legitimizing the property of people who accumulated such even through foul means during the heyday of colonialism and feudalism. The issue of land rights for the landless or "asset-less" people is neither thought of in the UN Declaration on Human Rights nor in ICESCR. Of late, the UN seems to have made an unsuccessful attempt to address the issue of peasants. In 2016, the Human Rights Council of the UN General Assembly constituted an open-ended intergovernmental working group on the rights of the peasants and other people working in the rural areas. The Chair-Rapporteur of the working group presented the 'Draft Declaration on the Rights of Peasants and Other People Working in Rural Areas' on 17-20 May 2016. It recognized that access to land, water, seeds, and other natural resources is an increasing challenge for peasants and rural people and stressed the importance of improving access to productive resources. It also underscored agrarian reform in national development strategies. Interestingly, this declaration defined peasant as "any woman or man who engages in or seeks to engage in small-scale agricultural production for subsistence/or for the market, and who relies significantly, though not necessarily exclusively, on family labor or household labor, and other non-monetized ways of organizing labor". The states have the obligations to respect, protect and fulfill the rights of peasants. It also refers to peasants as equal to all other people. Specification is also on rights to sovereignty over natural resources, development and food sovereignty, right to land, and other natural resources, and right to means of production, seeds, biological diversity, water, and housing. This is indeed a progressive draft document of the UN but unfortunately, it was only discussed in the intergovernmental working group and could not be passed and therefore, it could not be sent to the Human Rights Council (Open-ended intergovernmental working group of the Human Rights Council on the rights of peasants and other people working in rural areas, 17-20 May 2016).

Place of the Rights on Lands in the Main Human Rights Standard Mechanisms

Given the fact the place of rights on land cannot be examined and analyzed in isolation, there is the need to consider it in the context of UN enforcement mechanisms of Economic, Social and Cultural Rights (ESCR). Literature shows that there are basically four main human rights standard mechanisms responsible for the enforcement of ESCR which automatically subsumes the 'rights on land'. These comprise as follows: (i) Committee on Economic, Social and Cultural Rights (CESCR); (ii) Optional Protocol to the International Covenant on Economic, Social and Cultural Rights (OP-ICESCR); (iii) Special Procedures (SPs), and (iii) Civil Society Monitors (CSMs).

CESCR is the body of experts that monitors the implementation of ICESCR by state parties. These state parties are expected to submit regular reports to the Committee on the implementation of ESC rights domestically. They have to report initially within two years of ratifying the Covenant and thereafter every five years. The Committee examines the reports and addresses its concerns and recommendations to the state party in the form of 'concluding observations'. OP-ICESCR entered into force on 5 May 2013. It allows the CESCR to receive and consider communications from individuals or groups who are victims of violations of any ESC rights of ICESCR under the jurisdiction of state party to the covenant. The Committee will only consider a communication after all domestic remedies have been exhausted, unless domestic remedies are unreasonably prolonged. Under SPs, the Human Rights Council appoints Special Rapporteurs or independent experts to address specific country situations or thematic issues. There are several thematic mandates which focus on ESC rights such as right to food, adequate standard of living, non-discrimination, access to resources, etc. NGOs and civil society groups are often effective actors in holding states accountable for human rights obligations. They put political pressures on states by making them aware that their actions are being watched. CSOs can engage in this process by lobbying their governments to ratify treaties, monitor states' compliance to 'concluding observations' and 'treaty obligations', submitting information to CESCR and participating in country review sessions as observers and through oral submissions (UN General Assembly, 1966, pp.23-24).

UN Special Rapporteurs' Highlights on Land Rights

The Special Rapporteur is an independent expert appointed by the Human Rights Council to examine and report back on a country situation or a specific human rights theme. This position is honorary and the expert is not a staff of the United Nations nor paid for his/her work. Since 1979, special mechanisms have been created by the United Nations to examine specific country situations or themes from a human rights perspective. The United Nations Commission on Human Rights, replaced by the Human Rights Council in June 2006, has mandated experts to study particular human rights issues. These experts constitute what are known as the United Nations human rights mechanisms or mandates, or the system of special procedures (www.org/EN/Issues/Food/Pages/FoodIndex/aspx downloaded on 7/27/2016). Literature search from early 2000 shows that instead of directly dwelling on land rights, UN Special Rapporteurs have been appointed on 'housing' (including adequate housing), 'right to food' and 'rights of indigenous peoples' (which have direct implications on land because of their inextricable link to land rights). Their highlights of the findings have been expressed underneath.

Rapporteurs on housing view the right to housing (which is contingent on the tenure of land) as an important facet of human rights because housing is the stability and security for an individual or family. As the centre of our social, emotional, and sometimes economic lives, a home should be a sanctuary; a place to live, a security, and the dignity. Increasingly viewed as a commodity, housing is most importantly a human right. Under international law, to be adequately housed means having secure tenure, not having to worry about being evicted or having your home or land taken away. It means living somewhere that is in keeping with your culture, and having access to appropriate services, schools, and employment. Too often violations of right to housing occur with impunity. In part, this is because at the domestic level, housing is rarely treated as a human right. The key to ensuring adequate housing is the implementation of this human right through government policies and programs, including housing strategies (www.ohchr.org/EN/Issues/Housing/Pages/HousingIndex.aspx. downloaded on 6/5/2016).

The development of human right to adequate housing as a legal and advocacy tool has gained momentum during the past decade, particularly through the consistent work of civil society. The related embedding of this right in the UN human rights program and as a key component of the Habitat Agenda has ensured the recognition of the right to housing as a cornerstone human right resulting in the recognition of the housing rights framework and the critical relevance of a gender and human rights perspective (in the context of global deterioration of housing rights for poor/vulnerable such as with no shelter at all and children in the street). Indeed, the 'right to adequate housing is the right of every woman, man, youth and child to gain and sustain a secure home, and community in which to live in peace and dignity' (a definition emanating from the national campaign for housing rights, the habitat international coalition, and committee on economic, social and cultural rights) (Report of the Special Rapporteur on adequate housing as a component of right to an adequate standard of living by Miloon Kothari, 25 January 2001).

There is indivisibility and interrelatedness of all human rights. In this context, there is the need of examination of a range of issues related to adequate housing (including gender discrimination, access to land, potable water and issues of economic globalization and its compatibility with human rights and particularly its impact on housing, the international cooperation dimension, forced evictions and poverty, and global social policies and their interface with human rights). Indeed, adequate housing is to be recognized as a distinct human right (Report of the Special Rapporteur on adequate housing as a component of right to an adequate standard of living by Miloon Kothari, 1 March 2002). In 2003, Special Rapporteur Miloon Kothari recommended to give the firm recognition to the human right to adequate housing and called for an expert group meeting to develop policy guidelines for preventing discrimination and segregation in housing and civil services (Report of the Special Rapporteur on adequate housing as a component of right to an adequate standard of living by Miloon Kothari, 3 March 2003).

Indeed, 'forced eviction' is to be the priority issue with respect to adequate housing as a component of the right to an adequate standard of living. The Commission on Human Rights adopted a resolution in 1993 recognizing 'forced eviction' as the 'gross violation of human rights'. Despite several UN and civil society efforts, 'forced evictions' still result in displacements, loss of livelihoods, property and belongings, and physical and psychological injury to those affected, which often include persons already living in extreme poverty, women, children, indigenous peoples, minorities and other groups at work. The Rapporteur analyzed several causes of evictions such as development-induced displacements, globalization (i.e. liberalization, privatization and growing land speculations), conflict and post-conflict situations, and punishment and use of excessive force and proposed range of measures to address it through the adoption of national policies and legislations and development of clear guidelines by experts to address evictions (Report of the Special Rapporteur on adequate housing as a component of right to an adequate standard of living by Miloon Kothari, 8 March 2004). The report of 2005 presented the analysis of the allegations of the rights violations and appeals to governments. In 2006, the Rapporteur confirmed the use of an approach stressing the indivisibility of human rights, without which the right to adequate housing loses its meaning. Whereas this approach meant focusing on the interface between adequate housing as an economic, social and cultural right and relevant civil and political rights such as the right to information and the right to the security of the home, his work has demonstrated that the existing inter-linkages go far beyond. An in-depth analysis of the multi-layered content of the human right to adequate housing necessitates the exploration of linkages with other related rights as the right to land, food, water, health, work, property, equality, inheritance and protection against inhuman and degrading treatment, with non-discrimination and security of tenure at the core. With this analysis, a recommendation was made to continue the mandate on adequate housing to initiate a process of adopting the guidelines on forced evictions and to consider recognition of land as a human right and, finally, a request was also made to arrest the urban and rural apartheid and segregation and to control unbridled property speculation and land confiscations (Report of the Special Rapporteur on adequate housing as a component of right to an adequate standard of living and the right to non-discrimination in this context by Miloon Kothari, 8 March, 2006).

In 2007, the Rapporteur provided practical and operational tools (e.g. indicators on the right to adequate housing, monitoring women's right to adequate housing, and land and basic principles and guidelines on development-based evictions and displacements) with a view to promoting, monitoring and implementing the human right to adequate housing. It also identified the normative gap – the non-recognition in international human rights law of the human right to land, underscored the use of gender perspective and cautioned against the practice of forced evictions (Report of the Special Rapporteur on adequate housing as a component of the right to an adequate standard of living by Miloon Kothari, 7 February 2007). In 2008, the Rapporteur confirmed the broad interpretation of the right to housing based on the indivisibility and universality of human rights and underlined the struggle against discrimination in the realization of right to housing and also advocated a humanitarian and human rights approach to address the situation of the people living in the grossly inadequate housing conditions and those facing homelessness and landlessness including support from the state (Report of the Special Rapporteur on adequate housing as a component of the right to an adequate standard of living and on the right to non-discrimination in this context by Miloon Kothari, 13 February 2008).

In 2009, in the context of the globalization of the housing and real estate markets and economic adjustment policies, the Rapporteur shared that cities have become unaffordable for inhabitants of lower income and increasingly middle-income groups (because of the growing nature of housing as a mere commodity and financial asset). The housing crisis was reflected upon because globalization accentuated the commodification of housing and state's role was recommended for intervention for the public housing of the poor (because of the failure of markets to provide housing for all) (Report of the Special Rapporteur on adequate housing as a component of the right to an adequate standard of living and on the right to non-discrimination in this context by Raquel Rolnik, 4 February 2009). The same Rapporteur worked in 2010 who focused on the need to integrate human rights standards (particularly the right to adequate housing in post-disaster and post-conflict reconstruction processes with focus on security of tenure). In 2011, she focused on the question of women and their adequate housing to ensure that women everywhere are able to enjoy this right in practice for which policy advancement was needed to address the issues related to inheritance, land and property (Reports of the Special Rapporteur on adequate housing as a component of the right to an adequate standard of living and on the right to non-discrimination in this context by Raquel Rolnik, December 2010 and 26 December 2011).

In 2012, the Rapporteur elaborated on the concept of security of tenure as a component of the right to adequate housing. The generally accepted definition of tenure reads, "Tenure is understood as the set of relationships with respect to housing, and land established through the statutory law or customary, informal or hybrid arrangements" but it is considered incomprehensive because it does not encompass the 'informal settlements' (of urban poor) which are self-made, spontaneous, self-managed and unplanned settlements and housing arrangements, initiated by the urban poor themselves who are generally characterized by precarious infrastructure and housing conditions (Report of the Special Rapporteur on adequate housing as a component of the right to an adequate standard of living and on the right to non-discrimination in this context by Raquel Rolnik, 24 December 2012). In 2013, the Rapporteur analyzed two alternative housing policies – rental and collective housing – that could play a key role in the promotion of enjoyment of the right to adequate housing for those living in poverty and called for a paradigm shift from the financialization of housing to human-rights based approach (Report of the Special Rapporteur on adequate housing as a component of the right to an adequate standard of living and on the right to non-discrimination in this context by Raquel Rolnik, 7 August 2013). In 2014, the Rapporteur analyzed the role of local and the sub-national levels of government through the consideration of how they can be fully engaged in the realization of the right to adequate housing. Emphasis has been laid on the decentralization of housing policy formulation and implementation guided by human rights so that local and sub-national governments have to be accountable to human rights obligations for which the state must ensure the capacity/resources needed to fulfill these obligations (Report of the Special Rapporteur on adequate housing as a component of the right to an adequate standard of living and on the right to non-discrimination in this context by Leilani Farha, 22 December 2014).

There has been the understanding that housing is more than a commodity and, therefore, five areas deserve critical considerations on urban rights agenda, namely, social exclusion, migration and displacement, persons and groups in vulnerable situations, land and inequality and informal settlements. There is a need for a new rights-based framework for urban law, policy and governance to eliminate homelessness and forced evictions. There is a need to consider a three-dimensional human rights definition of homelessness: (i) the first dimension addresses the absence of home in terms of both its physical structure and its social aspects; (ii) the second dimension considers homelessness as a form of systematic discrimination and social exclusion, whereby "the homeless" become a social group subject to stigmatization, and (iii) the third dimension recognizes homeless people as resilient in the struggle for survival and dignity and potential agents for rights holders. And, therefore, the governments must have the human rights framework to address homelessness because their policies are bereft of human rights perspective (Human Rights Office of the High Commissioner, Report.19 and 20 Info Notes, 2014 based on the summary of the report of the Special Rapporteur Leilani Farha on the right to adequate housing).

Given the fact that 'the right to food' is ensured only with the access to land (a means of producing the food) and, therefore, a brief discussion on the role of the Special Rapporteur on the right to food is contextual in this paper. The 'right to food' is the right to have regular, permanent and unrestricted access, either directly or by means of financial purchases, to quantitatively and qualitatively adequate and sufficient food corresponding to the cultural traditions of the people to which the consumer belongs, and which ensure a physical and mental, individual and collective, fulfilling and dignified life free of fear. This definition is in line with the core elements of the right to food as defined by <u>General Comment No. 12</u> of the United Nations Committee on Economic, Social and Cultural Rights (the body in charge of monitoring the implementation of the ICESCRs in those states which are party to it). The Committee declared that "The right to adequate food is realized when every man, woman and child, alone or in community with others, has physical and economic access at all times to adequate food or means for its procurement (such as land). The right to adequate food shall, therefore, not be interpreted in a narrow or restrictive sense which equates it with a minimum package of calories, proteins and other specific nutrients. The right to adequate food will have to be realized progressively. However, states have a core obligation to take the necessary action to mitigate and alleviate hunger even in times of natural or other disasters. The nature of the legal obligations of state parties is set out in Article 2 of ICESCR. The Committee on Economic, Social and Cultural Rights in General Comment No. 12 also defined the obligations that state parties have to fulfill in order to implement the right to adequate food at the national level. These are as follows: (i) the obligation to respect existing access to adequate food requires state parties not to take any measures that result in preventing such access; (ii) the obligation to protect requires measures by the state to ensure that enterprises or individuals do not deprive individuals of their access to adequate food; (iii) the obligation to fulfill (facilitate) means the state must pro-actively engage in activities intended to strengthen people's access to and utilization of resources and means to ensure their livelihood, including food security, and (iv) whenever an individual or group is unable, for reasons beyond their control, to enjoy the right to adequate food by the means at their disposal, states have the obligation to fulfill (provide) that right directly. This obligation also applies for persons who are victims of natural or other disasters. While all the rights under the Covenant are meant to be achieved through progressive realization, states have some minimum core obligations which are of immediate effect. They have the obligation to refrain from any discrimination in access to food as well as to means and entitlements for its procurement, on the grounds of race, colour, sex, language, age, religion, political or other opinion, national or social origin, property, birth or other status. Finally, states have to ensure the satisfaction of the minimum essential level required to be free from hunger. The mandate of the Special Rapporteur on the right to food was originally established by the Commission on Human Rights in April 2000 by Resolution 2000/10. Subsequent to the replacement of the Commission by the Human Rights Council in June 2006, the mandate was endorsed and extended by the Human Rights Council by its Resolution 6/2 of 27 September 2007. The Special Rapporteur implements the mandate by reporting to the Human Rights Council and to the General Assembly on the activities and studies undertaken in the view of the implementation of the mandate, monitoring the situation of the right to food throughout the world, identifying general trends related to the right to food and undertake country visits which provide a first-hand account on the situation concerning the right to food in a specific country, and communicating with states and other concerned parties with regard to alleged cases of violations of the right to food and other issues related to their mandates. (www.org/EN/Issues/Food/Pages/FoodIndex/aspx downloaded on 7/27/2016).

It would be contextual to have a brief discussion on the different foci of the Special Rapporteurs in a period of 15 years (from 2001 to 2016) and to see the connectivity between 'the right to food' and 'land rights'. More specifically, in 2010, the focus was on access to land and the right to food, agri-business (including the addendum on the analysis of large-scale acquisitions and leases, their negative impacts on the livelihood of the smallholders and socially and economically vulnerable people). In 2011, a discussion used human rights criteria for making contract farming and their business models inclusive of small-scale farmers (including the focus on agro-ecology).

It is also relevant to have a brief discussion on the contribution of the UN Special Rapporteur on the rights of indigenous peoples (IPs) vis-à-vis land rights. Victoria Tauli Corpuz (a Filipino lady) was appointed in 2014 and prepared a report on the impact of international investment and free trade on IPs' rights which was presented to the 70th Session of UN General Assembly in 2015. She contends that the investment clauses of the free trade and bilateral and multilateral investment treaties, as they are currently conceptualized and implemented, have actual and potential negative impacts on IPs's rights, in particular on their rights to self-determination, lands, territories and resources, and free and informed consent. That is not to suggest that investment agreements can be equally beneficial for IPs and investors. She contends that the right to free, informed, and prior consent is included within the UN declaration on the rights of IPs and the right to consultation in ILO Convention No. 169. Despite these provisions, representatives from national governments negotiate, draft and agree on investment agreements in strict privacy devoid of the representation of IPs. The cultural rights of IPs are constantly undermined by effects of investments and free trade agreements. Their cultural rights are linked to land and waters which are again integral to their culture and identity. Therefore, barriers to indigenous land ownership created by international investment agreements/free trade agreements are also assaults on the cultural rights of IPs because the displacement commonly caused by the loss of land and territory can further undermine the cultural integrity and protection of indigenous communities. The UN Declaration (Article 32, Para 1 states that "Indigenous peoples have the right to determine and develop priorities and strategies for the development or use of their lands or territories". But international investments and free trade agreements trigger gross and sustained assaults on the cultural integrity of IPs, denigration and non-recognition of customary laws and governance systems vis-à-vis the land and natural resources resulting in their economic deprivation. The key challenge has been the dominance of neo-liberalism (which champions the power of market forces) and its focus on extractive activities by the vested interests with the exploitation of natural resources (e.g. minerals, metals, oil, gas, water, and products from forests, farming and fishing) which have negative bearing on indigenous peoples. There is always the need for a participatory national mechanism to allow them to take part in the negotiation and drafting of such agreements. There must be social dialogues of IPs with all levels of governments and there must be the human rights assessments prior to the signing of all such treaties. Definitely, the governments must consider protecting rights of IPs (Corpuz, 1015). But many of these considerations are equally applicable in the national development project activities vis-à-vis the land rights of IPs.

In 2012, the Coalition on World Food Security (CFS) and UN Food and Agriculture Organization (FAO) has developed Voluntary Guidelines on the Responsible Governance of Tenure of Land, Fisheries and Forests in the Context of National Food Security (2012) which discusses the guiding principles of responsible tenure governance (including rights and responsibilities related to tenure), legal recognition and allocation of tenure rights (with focus on safeguards, public land, fisheries, and forests, indigenous peoples, and other communities with customary tenure systems and informal tenure), and transfers and other changes to tenure rights and duties (with focus on markets, investments, land consolidation, restitution, redistributive reforms, and expropriation and compensation). CFS and FAO (2012) hold that the purpose of these Voluntary Guidelines is to serve as a reference and to provide guidance to improve the governance of tenure of land, fisheries and forests with the overarching goal of achieving food security for all and to support the progressive realization of the right to adequate food in the context of national food security. These guidelines are intended to contribute to the global and national efforts towards the eradication of hunger and poverty, based on the principles of sustainable development and with the recognition of the centrality of land to development by promoting secure tenure rights and equitable access to land, fisheries and forests (given the fact that the eradication of hunger and poverty, and the sustainable use of the environment, depend in large measure on how people, communities and others gain access to land, fisheries and forests). The livelihoods of many, particularly the rural poor, are based on secure and equitable access to and control over these resources. While recognizing the existence of different models and systems of governance of these natural resources under national contexts, states may wish to take the governance of these associated natural resources into account in their implementation of these Guidelines, as appropriate.

How people, communities and others gain access to land, fisheries and forests is defined and regulated by societies through systems of tenure. Inadequate and insecure tenure rights increase vulnerability, hunger and poverty, and can lead to conflict and environmental degradation when competing users fight for control of these resources. The governance of tenure is a crucial element in determining if and how people, communities and others are able to acquire rights, and associated duties, to use and control land, fisheries and forests. Many tenure problems arise because of weak governance, and attempts to address tenure problems are affected by the quality of governance. Weak governance adversely affects social stability, sustainable use of the environment, investment and economic growth. Responsible governance of tenure conversely promotes sustainable social and economic development that can help eradicate poverty and food insecurity, and encourages responsible investment (CFS and FAO (2012, pp.iv-v). The document of these guidelines is of progressive nature in the sense that it has analyzed the role of safeguards, customary tenure systems and even the informal tenure within the regime of responsible governance that definitely create an ambience for the land resource-poor to have the realization of their land rights. But the problem is that the states may or may not comply with these voluntary guidelines.

Land Rights Problems in South Asia as Human Rights Problems

Land Rights Problems in Nepal

An attempt has been made to analyze the land rights problems based on a number of issues. These subsume: political economy of land rights problem, contemporary context of land rights, women's issue on land, and the need to implement land policies addressing the concerns of rights holders (addressed by recent constitution and lands act amendments).

Political economy of land rights problem in Nepal in a nutshell

The main objectives of the institutional land policy of the rulers after the political unification of Nepal in 1768 were to administer areas, make land grants and assignments to different categories of people close to power structures and extract revenues directly or indirectly to further the objective of territorial expansion (Regmi, 1971, p.37). Successive governments adopted a policy to grant *Birta* (tax-free), *Jagir* (land assignments to the civil and military officials in lieu of emoluments) and *Guthi* (land assignments to the temples, monasteries, educational institutions, etc for philanthropic and religious purposes). Indeed, these land assignments were made from the *Raikar* (state-owned taxable land which also included virgin land for reclamation). These land tenure systems were predominantly practised by the hereditary autocracy from 1846 to 1951. Then, state land (mostly virgin) was again distributed to the loyal supporters of the *Panchayat* (a party-less dictatorial political system under absolute monarchy) introduced in 1960 by banning the multi-party political system introduced in 1951 due to the political movement. These land tenure systems created a highly inegalitarian society in which the generality of the peasantry had to survive on the mercy of the feudal landlords and the ecclesiastics of the temples and monasteries by working in their farmlands with high agricultural rents (to be paid). The *Kipat* (communal land ownership) tenure customarily practised by the *Mongoloid* indigenous groups of eastern Nepal were also legally abolished in 1968 in the process of implementing the 1964 Lands Act so that the state could appropriate revenue from such land with the use of progressive taxation policy (because tax on *Kipat* land was imposed on homestead only irrespective of the size of the cultivated land owned by a household).

Land reform policies of the past and their fiasco

In 1959, the first ever democratically elected government of Nepal enacted the *Birta* Abolition Act guided by the general egalitarian principles.

However, the total elimination of practices in the nook and corner of the country has not been fully realized hitherto. The 1964 Lands Act under the monarchy had made some effort for the land reform in the country (which was largely under the influence of the foreign donors, principally by the USA which had been advocating for the land reform to deter communism). This Lands Act set the land ceilings. For instance, ceilings for the landowners were set at 2.7 hectares in the Kathmandu valley, 4.1 hectares in the Hills and 17 hectares in the *Tarai* whereas land ceilings were fixed for land rented/leased by tenants at 0.5, 1.1 and 2.7 hectares in the Kathmandu valley, the Hills and the *Tarai*, respectively. These above data clearly show that the law favored landowners over tenants. In fact, landowners had a much greater advantage over tenants in terms of the size of land ceiling and production potential in all parts of the country. For instance, in addition to the land permitted under the ceilings, landowners were allowed to retain homesteads, whereas no such provisions were made for tenants. Moreover, as sons above 16 years of age and unmarried daughters above 35 years of age were recognized as separate families for homestead purposes (even if they operated holdings together), it was possible for landowners to retain considerable proportions of land as ostensible homesteads.

Not surprisingly, the outcome of the reform measures in terms of the appropriation and redistribution of excess land was unsuccessful. By 1978, about 50,000 hectares were declared as access land, out of which 34,705 hectares was actually confiscated; of this confiscated land only 21,050 hectares were redistributed. Moreover, it appears that a significant portion of this redistributed land was procured by government officials and their families, relatives and friends rather than tenants and landless. Similarly, inequalities continued in the rates of agricultural rents paid by the tenants. The 1964 Act set land rents at 50 percent of the gross product (later amended to 50 percent of the main crop) and tenants were required to bear the costs of cultivation. This share-cropping arrangement was, in fact, the same as before the introduction of land reform program. The land legislation measures resulted not only in the continuing economic and tenurial insecurity of tenants but in many cases, they resulted in the complete termination of their access to land. The absence of awareness about the reform measures amongst the tenants and the landless and the lack of real political will and determination on the part of the ruling elite involved in the reform measures meant that no structural changes within the existing agricultural relations were achieved; the poorest were increasingly rendered more vulnerable to loss of land rights (Ghimire, 1998, pp. 39-41).

Besides the above, the 1964 land reform measure was implemented in three phases. The implementation began from the eastern Nepal where the concentration of land in the hands of landlords was less as compared to the western, mid-western and far-western regions (particularly in the *Tarai* region). This government's deliberate intention for the implementation gave the landowners who were anticipating land transfers to skirt the law by putting their land in the names of their relatives, loyalists, favorites, and friends, triggering the total failure of expropriating the excess land above ceilings (Zaman, 1973).

Contemporary context of land rights problem

In Nepal, land ownership remains the main source of wealth, social status, and economic power in the agrarian social structure of the country and, therefore, the land rights movement program is of paramount importance – a function of the inequity in the land resource distribution. It is axiomatic that social and economic transformation remains incomplete in Nepal in the absence of land reform. Social transformation is impossible without the transformation of power relationships (even if the state restructuring takes place in the days to come). There is a need to transform the existing land and power relationships due to the following factors:

(a) **Need to address the issues of landlessness:** The Nepal Living Standards Survey (NLSS) of 2010/11 also reveals the limited landholding in Nepal, a major trigger for the incidence of poverty in rural Nepal. More specifically, a sizable proportion of people (44 percent) have land between 0.2-1 hectare followed by 21 percent landless. Similarly, less than one-fifth of the population (17 percent) have less than 0.2 hectare followed by 13 percent having land between 1-2 hectares and 8 percent having land more than 2 hectares. Poverty and size of landholding is positively correlated. The headcount rate of population falling below the poverty line is the highest (29.9 percent) with land size less than 0.2 hectare followed by 28.2 percent with land size 0.2-1 hectare, 22.7 percent among the landless, 19.1 percent with land size 1-2 hectares and 6.5 percent with land size more than 2 hectares. To a large measure, the incidence of poverty has been found to be less among households having more than 1 hectare of arable land. The larger the size of the landholding is, the lesser the incidence of poverty is found (CBS, 2011). In 2014, by doing the re-analysis of the 2011/12 agricultural census data, Gefont shows that there are 9,33,044 agricultural households in Nepal which are landless and poor households with land less than 0.2 hectare. This constitutes 24 percent of the total 38,85,093 agricultural households of Nepal (Gefont, 2014). Therefore, there is the urgent need to continue movement and policy advocacy to address the issue of landlessness.

(b) **Need to pressurize the government to implement the constitutional provisions vis-a-vis land rights**: Given the context of the recent promulgation of the new constitution of the federal republic of Nepal in 2015, it is worthwhile to visit it by examining a few important issues related to land rights and agrarian reform addressed by the constitution which were communicated to the constituent assembly (CA) by the Community Self-Reliance Centre (CSRC) and its partner National Land Rights Forum (NLRF) because these can be labelled as the 'tectonic shifts' in the constitutional history of Nepal initiated largely for the greater social equity and social justice among the historically and structurally marginalized/excluded segments of Nepalese society (for which there have been unremitting struggles for more than 65 years after the democratic emergence in 1951 and the CSRC and NLRF have also made significant contributions to this domain). Massive seamless campaigns from the grassroots to the policy level people by the NLRF and the CSRC for seven years since the time of the election of first CA in 1966 to the time of final drafting of the new constitution by the second CA in 2015, and the holding of innumerable meetings with the main leadership of all major and minor political parties represented in the CA and members of the CA can be attributed to these 'tectonic shifts' on issues related to land rights and agrarian reform. Indeed, this has always been the campaign of the CSRC and the NLRF. Nonetheless, this outcome cannot be attributed to their campaign alone. The contribution of other women rights activists and political party activists is equally important. The sub-article two of Article 26 of the new constitution on 'Right to Religious Freedom' states, "Each religious community shall enjoy the right to religious freedom and operate and preserve the religious trust". It further states, "But it shall not be considered an obstruction for the regulation through the formulation of necessary laws to operate and preserve the religious site and religious trust and the management of trust property and land". This provision largely reflects the suggestion of the NLRF/CSRC which is possible through the amendment of law germane to trust land. Article 34 (sub-article 2) on 'Right to Labour' ensures the fair wages which has been the part of the CSRC/NLRF campaign. 'Right to Housing' has always been the slogan of the NLRF with the support of the CSRC and in this context, Article 37 (sub-article 1) on 'Right to Housing' states, "Each citizen shall have right to appropriate housing". Article 40 (sub-article 5) '*Dalit*'s Right' states, "The state shall make land available to landless *Dalits* once as per the law" and sub-article (6) says, "The state shall manage the settlement for *Dalits* deprived of housing as per the law" (CSRC/NLRF's suggestions largely accepted). Given the fact that both NLRF and CSRC work for gender equality (as indicated above), 'Women's Right' under Article 38 (sub-article 6) is of great interest. It states, "Couple shall have equal right to property and familial affairs". Again, this has been the function of the contribution of women rights activists of the CSOs and political party activists including that of advocacy campaigns of the NLRF/CSRC. Article 42 (sub-article 4) on 'Right to Social Justice' states, "Each farmer shall have the right to have access to land for agricultural work, select and preserve local seeds and agricultural species traditionally used and adapted". What is important is the emphasis on farmers' access to land for agricultural land – a main agendum or slogan of the NLRF/CSRC campaign right from the very beginning. Now it is worthwhile to review the 'E' sub-section 'Policy Concerning Agriculture and Land Reform' in Article 51 under part four of the constitution entitled 'Directive Principles, Policies and Responsibilities'. It has five constitutional clauses as follows: (i) 'scientific land reform to be implemented by paying attention to the interest of farmers through the abolition of the existing dual ownership on land (NLRF/CSRC suggestion spirit included); (ii) 'augment production and productivity by consolidating the land through the discouragement of absentee landlordism (NLRF/CSRC suggestion spirit included); (iii) land management and agricultural professionalization, industrialization, diversification and modernization to be done through the adoption of land use policy for the augmentation of agricultural production and productivity by protecting and promoting the rights and interests of farmers (NLRF/CSRC suggestion spirit included); (iv) integrated utilization of land through regulation and management on the basis of productivity, nature and environmental balance (for the materialization of which the NLRF and CSRC have been campaigning for long for the endorsement of national land policy by the cabinet), and (v) management for farmers to be ensured on agricultural inputs, reasonable pricing of agricultural products and access to markets (for which CSRC has been working to promote the agricultural cooperatives as a part of land rights and agrarian reform). Similarly, in the 'I' sub-section of the same 'Policy', there is a constitutional clause which states, "rehabilitation of freed-bonded labourers, ploughmen, cattle-herders and landless *Sukumbasis* (squatters) through their identification by managing homestead land for housing and agricultural land or employment for livelihood" (this has been always the agenda of the NLRF/CSRC advocacy campaign to address the issues of landless agricultural labourers – the goal for empowerment (Uprety, 2015). These constitutional provisions enshrined in the constitution will be translated into relevant acts and policies and implemented as per the constitutional spirit provided regular advocacy and campaigns are launched with the participation of the primary stakeholders.

(c) **Need to pressurize the government to implement the 6ᵗʰ amendment of 1964 Lands Act for settling the tenancy-related issues**: The 6th amendment to the 1964 Lands Act for settling the tenancy-related issues was passed in the Legislature-Parliament on the 6th of September, 2015. Pursuant to the agreement concluded between the NLRF and the Ministry of Land Reform and Management (MoLRM), the amended Act incorporates the provisions for keeping the door open for settling the tenancy cases, and permitting the Land Revenue Offices to deploy the team of officials in the villages and settle the tenancy cases on the spot instead of requiring the affected parties to come to it at the district headquarters. This also requires continuous campaign for putting pressure on the government.

(d) **Need of drafting new land acts**: There is a need to continue the process of drafting the new *Birta* Abolition Act and *Guthi* Land Management Act in close consultation with the NLRF for the benefit of actual tillers through registration in their names.

(e) **Need to address the gender inequality vis-à-vis land ownership**: There is increasing feminization of agriculture in Nepal. Women traditionally constitute a sizable proportion of agricultural labour force in the agrarian economy of rural Nepal. For instance, the Nepal Living Standards Survey (NLSS) of 2010/11 conducted by the Central Bureau of Statistics (CBS) shows that 32 percent of males have been involved in non-agricultural productive activities whereas female involvement is only 11 percent, demonstrating their much higher involvement in the agriculture sector. Not surprisingly, the ever-increasing trend of out-migration of Nepali male labourers to East Asian and the Gulf countries as well as India for the jobs (a function of a multitude of factors such as political instability, insecurity, unemployment, underemployment, poverty, lack of job opportunities, etc) has played a pivotal role in feminizing the agriculture of Nepal which, in turn, has increased women's roles and responsibilities in the productive sector (besides their traditional reproductive roles) for taking care of multifarious activities in their traditional subsistence farms. These comprise production, processing and preparation of the food, *inter alia*. Succinctly put, they have played an instrumental role in ensuring food security for the households and communities. The ever-increasing feminization of agriculture is reflected in the demographic composition of female-headed households in Nepal. Apropos of it, being based on 2011 decennial census data, the CBS in 2012 states that female-headed households in the country had increased by about 11 percent from 14.87 percent in 2001 to 25.73 percent in 2011. The irony is that despite the laudable contribution made to the agricultural sector of Nepal, their marginalization continues to be unabated and their identity as farmers has not been established and recognized. This is manifested in the lack of ownership and control over productive resources (principally land) and outputs. The national data made available by CBS in 2012 show the dismal picture of female ownership of fixed assets. CBS states that during the decennial census of 2011, altogether 19.71 percent of households reported the ownership of land or house or both in the name of the female member of the household. In urban areas, 26.77 percent of the households show female ownership of fixed assets while the percentage stands at 18.02 in rural areas (CBS, 2012, p.2).

Review of Landesa's study in the form of briefing paper (2015) for Nepal also helps generate a lot of information regarding the status of women's land rights in Nepal. The key findings of the paper include: (i) there are several opportunities for women to own land in Nepal. For instance, equal inheritance and progressive partition provisions in the *Muluki Ain*-civil code (as amended in 2002) and government policy for tax rebates and joint title ownership in favour of women. Despite these opportunities, women do not have equal rights to land ownership and control in practice in Nepal – a function of the culture of patriarchy. As indicated above, around 19.71 percent of women in Nepal own land or house or both as per the CBS census data of 2011. But an ANGOC's study on 'women's land rights in Asia' quotes a study by the Ministry of Health and Population and its associates (2012) that only 9.7 percent Nepali women owned land in 2011. The prime barriers in women's access to land rights include: lack of awareness about inheritance to land rights among women, revenue officials and CSOs; improper and inadequate implementation of the *Mulaki Ain* (Civil Code); patriarchal attitude of community members, revenue officials and individuals towards women; lack of gender friendly policies in related institutions; inadequacy of institutions to deal with the issue of landlessness; lack of monitoring and revision of laws and programs in favour of women; policies and programs not reaching out to the real beneficiaries (i.e. rural poor *Dalit* women); women's lack of access to land-related institutions, etc. Landesa (2015) in Nepal also dwells on the regulatory aspect in a greater detail. For instance, the Nepal-specific paper asserts that there are no clear instructions on joint title under government land allocation scheme but women, after labour movement, are getting land under joint title under the Freed Bonded Labourers' Scheme. The government has initiated a scheme called *Sayamukta Lal Purja* (joint ownership certificate) to promote joint ownership of property announced in the Budget Speech of 2010/11 through easy and subsidized procedure (i.e. registration fee is Rs.100 only and today the tax exemption is 25 percent for land in urban areas, 30 percent for land in rural areas, 40 percent for land in the mountain region, and 30 percent when the woman is single) and this has been achieved after persistent efforts and advocacy by CSOs (such as CSRC and NLRF) and women groups by mobilizing urban and rural women. Positionalities of women are critical variables in affecting their legal right to own land. For instance, unmarried daughters have equal co-parcenary rights like that of son in parental property as per *Mulaki Ain* amended in 2002 (but married daughters are excluded from receiving parental property despite the constitutional guarantee of daughter's equal right to ancestral property). There is certain progressive provision in the legal code which enables married women to seek a share in the husband property even after the divorce. There is a provision for widows, daughters and sisters who can receive 'Juini' (a share in the ancestral property conferred on a person for her maintenance) so that she can be made secure in old age. The consolidated study paper of Landesa (2015,pp.14-17) writes that even through there are legal provisions in Nepal for women's land rights, there are a number of gaps. For instance, the issue of joint titles is problematic because it does not lead to absolute control of women over land unlike sole titles. Like in India and Pakistan, the core issue is inheritance which is a legal right whereas joint ownership is a policy/program which may be altered any time and joint ownership benefits only a section of women (such as the married ones and this has wider implications on women). As in other countries, the execution of government program is impeded due to the problem of identification of beneficiaries, lack of awareness among beneficiary women in remote hinterlands, and the patriarchal attitude of government revenue officials (which hampers the design and execution of these schemes). Traditionally, women are not considered as farmers in state policies. In the Nepalese context, there is deliberation emerging among civil societies to redefine "farmer" considering the fact that going by the slogan "land to the tiller", women cannot be considered as "farmers" as they are not allowed to till land owing to the cultural taboos. Apropos of gaps in legal rights, there are discriminatory positions and practices on the inheritance rights because women are not equal to their male counterparts. The Nepal-specific briefing paper (2015,p.15) asserts that there is *de jure* equality in inheritance rights but lack of *de facto* implementation. The Nepal-specific briefing paper of Landesa (2015. pp.14-28) reports that the impact of tax rebates on women in joint land ownership has been limited. There is also limited knowledge among women about the joint ownership scheme. More women in urban areas and upper/middle have the probability of being benefitted from the scheme. Although joint ownership is a good initiative, it does not provide absolute control over land as the sole title does. Despite the constitutional guarantee and declaration by the House of

Representatives on 31 May 2006 to issue citizenship in the name of a mother (a must for enjoying property rights in Nepal), there has been hardly any change in the practice due to the patriarchal mindset among the bureaucrats. There is also a need to redefine the notion of 'farmer' (i.e. although women are socially recognized as farmers, culturally they are not because it is taboo for them to till and at times the absentee landlords are considered as the farmers). The Landesa briefing paper on the status of women's land rights in Nepal (2015,p.33) has also recommended key actions as follows: (i) training and gender sensitization of local government body officials and district revenue officials; (ii) spreading legal literacy among women in order to widen the reach and scope of women land rights; (iii) political parties need to take the lead in doing away with deficiencies in the existing law such as allowing married daughters' a share in parental land (which is now rectified by the current constitution of 2015 with the emphasis of equal right of daughters to the parental property); (iv) enforcing *Muluki Ain* especially the provisions affecting women's land rights (which is being revised in Nepal upon the proclamation of the new constitution in 2015); (v) different categories of tax rebate for different sections of women depending on the index of their socio-economic vulnerability with higher rebate for *Dalit* and rural landless women; (vi) need for strategic research on this issue for robust evidence-based advocacy; (vii) opening up the registration of new tenants including for women, and (viii) women land rights need to be taken up as a socio-political cultural campaign by multi-stakeholders for it to be an effective reality. All this justifies the continuity of the concerted advocacy and campaigns for the realization of women's land rights.

(f) **Need to address a host of other land rights-related issues**: These would include the need for the implementation of a national land policy (drafted) and the land use policy (approved); the problems of good governance in land administration (due to the excessive centralization of power and corruption); the exploitative socio-economic system (despite legal/policy prohibition on the use of bonded-ploughmen in mid- and far-west Nepal, bonded cattle herders in Siraha and Saptari of eastern *Tarai*, their rehabilitation has been a far-fetched dream); problem of unregistered tenants under the *Ukhada* (under which cash land tax used to be paid); the problem of sustained and improved livelihood of the land-poor; the influence of globalization (i.e. land-grabbing for commodification and the loss of land by actual tillers), and the lack of political commitment ("whose land? "Tillers"' is limited in the slogans of political parties since 1951). Scientific land reform is needed in the country to ensure peace and stability through the equitable redistribution of the landed resources, benefitting the poorer and marginalized sections of society. The movement for land rights continues to be a social movement against injustice, exploitation and poverty in the semi-feudal Nepali society which is possible only through the transformation of existing unequal and inequitable power relationships. Thus, the land rights movement is very relevant in the context of Nepal and will continue to be so.

Land Rights Problems in Bangladesh

The context

The noted publication entitled 'Land Governance in Asia: Understanding Debates on Land Tenure Rights and Land Reform in Asian Context' (2013) by Antonio B. Quizon, an eminent land rights activist, is worth reviewing to contextualize the discussion on the land rights problems in Bangladesh. He refers to the East Bengal State Acquisition and Tenancy Act of 1950 which played an instrumental role in the abolition of the *Zamandari* system (a system under which local landlords were appointed by the state authority for settling people in lands, collecting revenue and handing it over to the state coffers after deducting minimum proceeds) instituted by the British colonizers and handing over the control of land to the actual tillers. But following the land reforms after the partition of India and Pakistan, a large number of Hindu *Zamindars* were forced to flee the territory of Bangladesh (then East Pakistan) and subsequently, the affluent Muslim peasants began taking up the role of money lenders and hold on the abandoned lands illegally. The 1950 Act which had set the land ceiling of 13 hectares of land per family was later changed to 50 hectares per family by the Karachi-based government in 1961. In the wake of the independence of Bangladesh in 1971, the government annulled the provision of land ceiling of 1961 and re-set the 13-hectare ceiling of 1950 with the pronouncement of Land Reform Policy in 1972. This policy provision also declared that all new diluvial and accreted lands would be regarded as *Khas* (public land) but the military coup of 1975 was a setback for the policy of land redistribution. In 1984, the revised Land Reform Policy reduced the ceiling to 8.1 hectares which failed to expropriate the additional 1 million hectares of surplus land. On the one hand, this policy prohibited the practice of transferring the land in another's name to circumvent land ceilings and, on the other, tenure security for share-croppers, established daily wage for agricultural laborers and set out share-cropping arrangements between land owners and tenants. The formulation of Land Reform Action Program in 1987 also further defined the poor people eligible for *Khas* land. On the one hand, the land reform initiative in Bangladesh remained an unfinished business and, on the other, much of the *Khas* land supposedly under government stewardship has been illegally occupied by rich peasants and as a corollary of it, land-related conflict cases are common, contributing to the clogging of judicial courts (Quizon, 2013, pp.29-30).

The protection of the right to land implies the protection of the right to basic necessities (such as the right to food, shelter and social security) and the emancipation of peasants and workers as guaranteed by the Constitution of the People's Republic of Bangladesh. Given the fact that every person in Bangladesh depends on land for shelter or livelihood either directly or indirectly, land right is a human right and every person has the right to own land because the constitution has guaranteed human right, principles of ownership and the right to property. But translating this into reality has been a far-fetched dream (Islam, et al., 2015).

Regarding the incidence of landlessness and inequality in land distribution, the statistical data are to be analyzed to contextualize the discussion. A total of 57 percent households in entire rural Bangladesh are landless (i.e. they do not own any cultivable land). Distribution of arable land is also extremely unequal. For entire rural Bangladesh, the Gini coefficients are 0.803 including the landless and 0.548 excluding the landless. In rural Bangladesh, about one-third of the farmers are pure tenants, that is, they do not own any cultivable land who have either share-cropping or cash-lease arrangements with landlords for the land they till. The distribution of operated land by farm size groups for entire rural Bangladesh is equally important for the analysis from the point of view of inequality. For instance, a total of 36.3 percent marginal farmers (with land below 0.5 acres) hold only 9.6 percent, 44.6 percent small farmers (with land between 0.5-1.49 acres) hold 37.8 percent, 11.8 percent medium farmers (with land between 1.5-2.49 acres) hold 21.6 percent and 7 percent large farmers (with land 2.5 + acres) hold 31.1 percent of the total operated land in the country (Integrated Food Policy Research Institute, 2013).

The nature of urbanization constitutes another significant dynamic of poverty and access to land. Marginal farmers and those rendered landless migrate to the cities in search of livelihood. Urban sprawl is consequently driving land prices up and increasing the incidence of land-grabbing. There is the preponderance of widespread land-grabbing victims among 32 different ethnic minority groups (1.2 percent of total households). About one million Hindu households have lost their 2.1 million acres (850,000 hectares) because of the enlistment of their property under the Vested Property Act (VPA) which continued as the Enemy Property Act (EPA) enacted during the Indo-Pakistan War in 1965. Of the total 12.1 million sharecroppers in 2009, less than one percent has legal documents. Poor fisherfolk's access to *Khas* is highly restricted affecting the livelihood of about 38 million people. At best, five percent of the total 830,356 acres (336 hectares) of *Khas*-water buddies available have been distributed among the poor, on lease basis. (Barkat, 2015,pp.54-55).

There are conflicting data on the limited land ownership of women in Bangladesh. Prof. Abdul Barkat holds that despite constituting half of the country's population, women have effective ownership of only four percent of land in rural areas as revealed by his research entitled "Assessing Inheritance Laws and their Impact in Bangladesh". His research shows that irrespective of the differences in terms of class, religion, and ethnic group, women are largely deprived of their justifiable rights to land (*Daily Star*, 5 May 2015). But referring to the study data analyzed by Keran, et al. in 2015, ANGOC shares that women in Bangladesh own 10.10 percent of land but the percentage of landowners who are women is 22.61 in 2012 (ANGOC, 2015). Women receive lower shares of land inheritance among both Muslim and Hindu communities. The Sharia law is not equal between males and females for ownership of land. According to the Sharia law, a female is entitled to only half of what a male receives from the parental property. But in practice, the female rarely gets what she is entitled to get due to patriarchal discrimination against women. Hindu women do not inherit land according to Hindu laws, which very often make their position in family and society very peripheral and vulnerable. Thus, Hindu family laws are discriminatory to women. Similarly, indigenous women face discrimination, social exclusion, structural marginalization, and systemic oppression in our patriarchal society. Descent and property are transmitted through male line in many indigenous communities where women remain systematically devalued without property and genealogical identity (Sourav, 2015). Finally, thus, the core issues of agrarian reform in Bangladesh are: (i) the distribution of *Khas* land among the poor and landless; (ii) the limited land rights of the religious and ethnic minorities; (iii) women's access to land, and (iv) the fishing communities' access to water bodies (Barkat, 2015).

Land Rights Problems in India

The Context

In 2007, Kamala Viswesworan critically analyzed the colonial legacies of the land rights issues in India. She is of the opinion that even the shortest amount of reflection on the notion of private property leads to colonial histories of land ownership that are contingent upon status and prestige. In independent India, it is difficult to answer the question, "Who owns the land?" without turning to the institution of "permanent settlement" introduced to India by the East India Company. Under the Mughals and in successor kingdoms, there was no real concept of private ownership of land. In practice, rights to land were shared among peasants, who enjoyed hereditary occupancy rights that often approached *de facto* ownership. The landed feudal lords who often served as village revenue officials had only limited power to alienate land from others. Lord Cornwallis's 1793 proclamation of "permanent settlement" required the landlords to pay the amount of fixed revenue to the state and gave them immunity against revenue increase in the state in addition to security of ownership. The permanent settlement thus declared landholders like the *zamindars*, *taluqdars* and *jagirdars*, as well as their subordinate intermediaries who were responsible for collecting taxes and revenues, to be land proprietors, dispossessing peasants of land rights in the process. In colonial India, the British developed three major forms of land tenure. The *Zamindari* or *jagirdari* tenure put into place by the permanent settlement covered 40 percent of land in Hyderabad State under the Nizam, as well as large portions of North and Eastern India, constituting 57 percent of all cultivable lands. *Ryotwari* tenure, introduced in the Madras Presidency in 1792, recognized *ryots* or individual cultivators as proprietors of land; it was widespread throughout South India, Maharasthra and Assam, and constituted 38 percent of all lands. However, even in the state of Hyderabad where 60 percent of the lands were under the *ryotwari* system, government middlemen responsible for revenue collection and awarded titles such as *desmukh* or *desai* also amassed large stretches of land, becoming powerful landlords in the process. The Telengana region of Andhra Pradesh bore the brunt of the abuse under both the *jagirdari* and *ryotwari* systems of tenure, though other regions of Andhra also suffered. A third type of land tenure called *mahalwari*, used villages as the collective units for revenue settlement, was well established in Punjab, Haryana, and Madhya Pradesh and accounted for 5 percent of all cultivable lands. Despite the existence of massive land reform legislation, these three land tenure systems are still in place in large parts of India, and have had long-term effects for agricultural development. The British legacy of land administration and its emphasis on the extraction of revenues, no matter the social, political or ecological conditions of the countryside, resulted in large-scale famines throughout the late 19th century, and the increasing destitution of those who worked the land but had diminishing rights to it. As one observer put it, "By the most charitable assessment, the British colonial government had perpetuated the impoverishment of India's population. Judged more rigorously, it had probably worsened their poverty (Visweswaran, 2007,pp.4-5).

The critical literature written by A. B. Quizon, the renowned land rights activist, in 2013 amply demonstrates that land reforms under India's federal system of government were initiated, legislated and implemented by each of the 15 states with necessary guidance from the central government. Starting in the 1950s, the states enacted legislation aiming at abolishing intermediary interests in land; regulating tenancy; setting land ceilings and distributing surplus lands above the ceilings; and redistributing public lands for agriculture and homesteads. The most notable land reform programs were implemented in the states of West Bengal and Kerala, especially during the rule of leftist parties, and in Uttar Pradesh immediately after independence in 1947. The first set of state laws involved the abolition of the *Zamindar* (intermediary rent-collector) and the parallel *ryotwari* (direct collection) systems, which were vestiges of British colonial rule and which at that time governed 95 percent of the country. State legislation gave these intermediaries proprietary rights only to that portion of the land under their cultivation and divested them of the remainder, although often with high levels of compensation. Under these acts, 20–25 million tenants were given proprietary rights over their cultivated lands and became landowners. The next set of laws sought to protect tenant farmers, who constituted more than one-third of all rural households and who worked under landlords without security of tenure. Almost all states passed tenancy laws that granted permanent rights to tenants and prohibited or regulated new tenancy arrangements. As a result, some 12.4 million tenants, or about 8 percent of India's rural households, gained land rights. However, tenancy reforms also led to large-scale evictions of tenants by landlords. All Indian states passed legislation on land ceilings that limited the amount of agricultural land that a family or individual could own. The laws authorized the government to take possession of lands in excess of the ceiling for redistribution to landless or land poor farmers. The laws on land ceilings differed between states in terms of where the ceilings were set (from 10 to 50 acres), the amount of land awarded to beneficiaries, and restrictions on beneficiaries transferring or selling the lands awarded to them. By the end of 2005, about 6.5 million acres of surplus land had been redistributed to 5.6 million households. This represented 1 percent of India's agricultural lands and 4 percent of rural households. Finally, some states allocated government land to land-poor families. These consisted of agricultural plots and homesteads, or housing plots. It is estimated that about 4 million people received home lots. Other land-related reforms followed in the 1980s: e.g. reforms ensuring women's land rights, legal aid and legal education, land purchase programs for home lots and gardens, and tenure reforms related to social and community forestry (Quizon, 2013,p.28).

Land issues in India

Socio-economic and caste census of India conducted in 2011 shows that 29.97 percent of households are landless (www.visionias.in, n.d). The Gini coefficient of the land cultivated by households for 2011-12 is 0.7204. There is a clear rise in inequality in distribution of land cultivated by households. In 2011-12, the top 10 percent of households cultivated about 50 percent of the land (Rawal, 2014). The share of the female operational land-holders is 12.79 percent as per the agricultural census of India conducted in 2010-2011 (Dhar, 2012) but the picture of the real entitlement looks bleak. These data on women's limited landownership are also closer to the findings of another study. It shares that Indian rural women owned 14 percent of land in 2011 (Swaminathan, et al, 2011 quoted by ANGOC, 2015).

Shivani Chaudhary, an eminent land rights activist scholar of India, in her presentation entitled, "Land Grabbing in India for Food Security" (which is undated) has raised a myriad of land issues in the Indian context. Discussing land distribution and ownership, she reveals some interesting facts as follows: (i) 60 percent of the cultivable land of India is owned by 10 percent of the population; (ii) unequal land ownership is the root cause of poverty; (iii) the population of landless and 'near-landless' is 220 million; (iv) 90 percent of landless poor are scheduled castes (*Dalits*) and scheduled tribes (indigenous peoples), and (v) the majority of poor work as agricultural laborers. She concludes that access to land is a key determinant of food security and livelihood protection. She further adds that rural women depend greatly on land for subsistence and perform more than 50 percent of all the agricultural work. Although 35 percent of the households in India are female-headed ones, less than 2 percent hold actual titles to land. Indeed, despite the increasing feminization of agriculture as a function of outmigration from the provinces, there is no due recognition of women as farmers. She also critiques the past land reform efforts/policy due to their failures to address the gender equity and redistribution of 2 million acres of land declared surplus (expropriated as being above the ceilings) primarily due to litigation and other associated reasons. She also shares that there is also the ubiquity of the phenomenon of 'land-grabbing' (taking of people's land by the state and non-state actors forcefully in the name of public purpose without following the due process, leaving no legal recourse). Key factors of it as analyzed by her India subsume development projects, special economic zones or SEZs (under which 1 million face eviction), and slum demolitions. She claims that India is estimated to have the highest number of people displaced annually as a result of 'development projects' in the world. To bolster this claim, she argues that 65 million people have been displaced ever since Indian independence in 1947 (which is alarming). She also critiques the neo-liberal economic policies and obsession with the growth of gross domestic product (GDP) which has been forcing changes in land laws to facilitate the conversion of agricultural land and easing of the land sale. Urban land-grabbing is also on the rise triggered by unplanned urbanization, rental of urban land, repeal of urban (ceiling and regulation) act, inequitable land use policies (triggering lack of space for urban poor in cities and towns), rampant real estate speculation, etc. (Chaudary, n.d).

Trade and investment agreements with foreign companies have been characterized by the investor protection, risk of community land being grabbed for large-scale investment, non-requirement of free and prior informed consent, longer-term lease (i.e. 99 years) with potential of interfering with land reform actions, etc. In the context of India, land-grabbing has a number of impacts. These would subsume: increased forced evictions (40-50 percent of displaced are tribal/indigenous peoples); increasing landlessness/homelessness; acute agrarian crisis; rise in farmers' suicide (more than 250,000 in 15 years which could be up to the period of 2012 as per her allusion); forced migration to urban areas and deepening poverty and hunger; arbitrary arrests/attacks/detentions of human rights defenders; criminalization of social movements, social unrest and violence (the rise of insurgency and counter-insurgency movements), disproportionately severe impacts on women, and violation of multiple rights on women (Chaudary, n.d).

As India rapidly industrializes, the government and private firms are seeking large tracts of farm land to build factories, power plants and highways, sparking off violent protests by farmers and others. Land has been a big issue in India. For many Indians, land is the only asset or social security that they possess and is a mark of social standing. Nearly 60 percent of India's 1.2 billion citizens depend on farming for a living and each hectare of farmland supports five people. Most development projects require huge amounts of land. For example, a proposed steel mill by South Korea's Posco in Orissa will be built on 1,600 hectares. A six-lane highway between Agra and New Delhi will require 43,000 hectares. Compensation ranges from between $4,300 a hectare, in the case of top steelmaker Arcelor Mittal's proposed plant of over 4,400 hectares in Jharkhand, to $14,600 per hectare, offered to farmers displaced by Posco's Orissa mill. Despite the seemingly attractive prices, farmers have few other livelihood options and land taken over for industrialization has been blamed for displacing hundreds of thousands of people. Protests against land being taken over have become more visible as the economy expands and the rich-poor gap widens (https://landwatchasia.wordpress.com/tag/land-rights-movement-in-India, downloaded on 7/19/2016). All this demonstrates that there is the need of social movement in India for the land rights of land-poor, women, *Dalits* and indigenous peoples.

Land Rights Problems in South Asia

A limited number of efforts have been made in the past to analyze the land rights problems in South Asia region as a whole. ANGOC seems to be in the lead in this regard since 2008. Assessing the land issues in South Asia in the context of the role of South Asian Association for Regional Cooperation (SAARC), it notes:

> SAARC's policy documents are replete with pronouncements on poverty alleviation, improving agricultural production and attaining food security. Poverty has been put at the centre and pro-poor strategies have been adopted pursuant to the call of independent South Asian Commission on Poverty Alleviation. In particular, SAARC's regional goal on livelihood ... defines the distribution of land to the landless in the region. SAARC's development goal on livelihood aimed to reduce by half the number of poor people by 2010. Two of the indicators under this target are: (i) proportion of population below the calorie-based food plus non-food poverty line, and (ii) distribution of state land to the landless tenants... (ANGOC, 2008, p.1).

However, ANGOC has critically assessed that this regional organizational mechanism fell short of providing the benchmarks and targets for land distribution. It has categorically stated that SAARC's declaration on land redistribution has been a mere 'lip service'. Although suggestions were made by the technical group working on livelihood to create assets for sustainable livelihood including natural capital (land and water), ANGOC demonstrates the ambiguity of SAARC's position on the importance of land rights as well as to the absence of an official declaration from SAARC on land rights and issues as they relate to the farmers in the region. SAARC has not recognized the inter-relatedness of poverty alleviation, agricultural production, food security and land rights/access to land even at the minimum (ANGOC, 2008, p.2).

ANGOC's literature shares that SAARC's social charter has failed to include the land rights issues confronted by a generality of poor rural farmers. However, apparently, SAARC seems to consider the rationale of land distribution to the sheer size of the landless people in the region with an embedded objective of accomplishing its development goal on livelihood (despite the recognition of the fact that there is continuing decline in the availability of land). However, food security has been mentioned in the SAARC's charter. The issue of food security can be linked to the land rights issues and ANGOC believes that this may serve as a powerful tool for advocacy on land rights and issues. There has been an awareness of land as a basic problem in South Asia but ambiguity reigns in this regard. Hence, ANGOC asks four fundamental questions: (i) whether access to land and land rights *per se* are considered main issues by the SAARC and its members?; (ii) how SAARC defines or perceives land issues?; (iii) what priority is given by the SAARC officials to the land issues?; and (iii) whether SAARC officials view land rights as an inter-related or separate issue, *inter alia* (ANGOC, 2008, 3). It is also conspicuous from the SAARC charter that the countries of the region are concerned with the poor without explicitly mentioning the poor farmers or land-based rural workers.

Although the technical committee on agricultural rural development works in the region, there is no unambiguous mention of a panoply of issues vis-à-vis land issues as follows: (i) land rights; (ii) agrarian reforms (redistributive policies); (iii) provision of access to productive resources such as land; (iv) tenurial rights; (v) sustainable use and management of common property resources – CPRs (forests, water, seeds, genetic resources, biodiversity, land, etc.), (vi) resettlement/relocation; (vii) access to legal instruments for land disputes; (viii) women's rights to land; (x) customary rights of IPs; (xi) stakeholders' participation in formulating agrarian reform policies; (xiii) monitoring committee on land-related issues, and (ix) agrarian reform in places of conflict/war (ANGOC, 2008).

The realization of equitable economic growth is impossible without the institutional effort for facilitating and ensuring the land-poor farmers to have access to land and land tenurial security. The SAARC goal of distributing land as poverty alleviation target and addressing food security remains unaccomplished without creating an environment for the landless farmers to be the owners of the land by addressing the security of the tenure.

Status of women's land rights in South Asian context

In early 2015, Landesa prepared a draft consolidated briefing paper entitled 'Status of Women's Land Rights in Three Countries of South Asia (India, Nepal, and Pakistan)' through a comprehensive study. This study commissioned by Oxfam as a part of Oxfam GROW campaign (an initiative seeking to promote better ways to grow, share and live together to help build a future where everyone on the planet always has enough to eat) aims at informing stakeholders on the struggles women are facing on land rights in South Asia (India, Nepal and Pakistan). In a nutshell, the main findings at the regional level are: (i) opportunities available to own land (given the fact that there are government land allocation programs where women are allotted land; there also exist other government level initiatives to promote land ownership among married women in specific countries, and women's land rights as daughters and widows are clearly laid out in inheritance laws); (ii) gaps in women's land rights (problems abound in design and implementation of government programs and schemes given the fact that single women are hardly ever taken into account; state policies seldom recognize women as farmers; practically legal rights remain unimplemented due to the preponderance of patriarchal socio-customary practices, and women legal rights yet to be asserted – a function of patriarchal attitude of society and officials), and (iii) efforts towards ensuring women's land rights (existent are the multi-stakeholder efforts at the level of government and civil society to ensure secure land rights of women; devising plans and policies to set the policy agenda for women land rights are subsumed in the governmental effort such as in the national plan documents; and mobilizing community members, creating women stakeholders by forming self-help groups, women co-operatives and women producer organizations enhancing women's agency through bridging the gap between farmer (women) and market, campaigning and advocacy on women land rights, maintaining partnership with governments and providing technical assistance to them, and turning law as a utilitarian instrument for achieving social change favoring women are subsumed under the civil society effort). Notwithstanding these positive points, various actions have been suggested as presented underneath: (i) the conduct of further research studies for generating further robust evidences on the issue of women's land rights; (ii) advocating with SAARC's technical committee on Women, Youth and Children to craft policies and programs in this direction; (iii) inclusion of women's land rights as prime goals in Sustainable Development Goals (SDGs) in the regional level advocacy by CSOs; (iv) initiating a regional declaration on the issue of secure land rights of women with the effort of civil society, government and people; (v) need to deliberate the issue of women's land rights at regional forums (such as One Billion Rising Campaign South Asia, South Asian Women's Network, and South Asian Foundation), and (vi) and potential of initiating multi-stakeholder advocacy at regional level (Landesa, 1915,pp.4-5). Definitely, Oxfam and its partners in South Asian region have steadily focused on women land ownership in smallholder agriculture program.

Succinctly put, the regionally undertaken study paper on women and land rights has been premised on the fact that ownership, access and control over land is potentially a contributing factor in altering the balance of power in gender relations (Landesa, 1915, pp. 9-10). Given the facts that women are denied the rights in ownership and control of land (Velayudhan, 2009), women's land ownership remains a crucial political arena of the process of women empowerment because land defines social status and political power in the village and structures relationships both within and outside the household (Agrawal, 1999 and 1996). Despite the fact that land ownership enhances the bargaining power of women which leads to decision-making within the household, in the community, and political society, they are particularly vulnerable because their land rights are obtained through kinship relationships with men or marriage. Women's land ownership improves food security and reduces poverty for the whole family (IFAD, 2012). There are social and economic benefits when women have secure land rights (Bourdreaux and Sacks, 2009). Given the fact that women's land ownership can help enhance household decision-making, social capital (i.e. education of theirs and their children), food security, sustainable employment opportunity, better yield and protection against HIV/AIDS, economic empowerment, participation in community decision-making and implementing activities and reduce domestic violence. Land ownership enhances women's social status, dignity, security, voice and freedom of a person or a family and it follows as corollary that women's landownership issue is the foundation upon which the struggle for ending the gender subordination can be based (Oxfam, 2013).

Land Rights Movements in South Asia: A Brief Analysis

This section contains a brief analysis of the land rights movements of Nepal, Bangladesh and India which are largely led by CSOs with the participation of land-poor farmers. CSOs have organized land-poor farmers, built their own community-based organizations (CBOs), trained the activists, developed the leadership capacity of land-poor farmers themselves, and provided the overall leadership for the movements in the mobilization of land-poor farmers and lobbying as well as influencing the policy-making and implementing processes at the macro, meso and micro levels.

Land rights movement in Nepal: Role of CSRC and NLRF as leading organizations

The Community Self-Reliance Centre (CSRC), a membership-based non-governmental organization (NGO), was founded in 1993 through the registration at the District Administration Office of Sindupalanchowk in central Nepal. It was initiated by the collective effort of a group of young and energetic school teachers with unwavering commitment to change the existing pattern of elite-dominated unequal and inequitable power relationships through the organization and mobilization of marginalized groups of people, especially tenants and landless farmers. It has been engaged in conscientizing and organizing land-poor farmers (agricultural laborers, tenants and marginalized farmers) who are deprived of their basic rights to land so that they can claim and exercise their rights over land resources in a peaceful way. The CSRC has adopted the human rights-based approach of development and its vision is 'a Nepali society where people have self-reliance and dignity'. Its mission is 'to enhance the power of land-poor farmers for leading land and agrarian reforms'. Its goal is 'to ensure land for land-poor farmers and their secure livelihood'. Its core values are: (i) simplicity of the lifestyle of its members and professional staff in their lives; (ii) belief in alternative solutions to every problem; (iii) people's struggle for their betterment with non-violent means; (v) emphasis on definite and tangible results with assured quality; (vi) compassion (sense of empathy with the people whose fundamental rights have been denied); (vii) taking sides (on the cause of land-poor farmers); (viii) self-reliance, and (x) sustainable land use. The guiding principles of the CSRC are: (i) commitment to the promotion and protection of all human rights; (ii) respect for plurality and diversity; (iii) social inclusion; (iv) non-violent action; (v) good governance; (vi) democratic decision-making and implementation, and (vii) adherence to the promotion of gender equality. The characteristics of the CSRC's theory of change are: (i) belief in land-poor people's ability and power to drive positive change; (ii) non-violent social movement; (iii) recognition of the instrumental role of elites and allies in setting public opinion and policy change, and (iv) production of knowledge as a precondition to design and win struggle. The strategic objective of the CSRC for July 2014-June 2019 as outlined in the strategy reads as follows, "The land and agrarian rights movement will strive to enable land-poor farmers (agricultural laborers, tenants, and marginalized farmers) to effectively use existing assets; maximize their potential; expand their opportunities to participate in decision-making that affects them; overcome isolating, discriminating or marginalizing; and work together to secure their land and agrarian rights" . The CSRC has been achieving this strategic objective through many strategies which include: (i) strengthening the organizational capacity of the NLRF and its local bodies/partners/units; (ii) enhancing food security and livelihood needs of land-poor farmers; (iii) promotion of non-violent and people-led campaigns; (iii) launching focused and coordinated movements complemented by concerted advocacy efforts; (iv) strengthening collaborative alliances with CSOs promoting human rights and facilitating movements for social justice; (v) working with policy think-tanks and academicians; (vi) enhancing women's leadership; (vii) developing women-led cooperatives and enterprises; (viii) expanding women's land ownership campaigns with different stakeholders; (ix) diversifying funds for mobilization and partnerships; (x) standardizing policies, systems and compliance, and (xi) generating, documenting and disseminating lessons. Realistically speaking, there is relatively a good manifestation of Nepal's social diversity in CSRC's organization. Despite the fact that the CSRC commenced its land and agrarian rights movement/ campaign from the two Village Development Committees (VDCs) of Sindupalanchowk District two decade ago, it has now been an organization of national repute primarily because of its movement/campaign expansion in 54 districts of Nepal with the objective of empowering land-poor farmers to claim and exercise their land rights at the community level. Interestingly, it has now the conviction that the movement/campaign has to be led by the *Pidit Pakchha* (affected party), viz; the land-poor farmers themselves and as a corollary of it, it has now turned into a resource center and a coordinating organization of the land and agrarian movement/ campaign which is being led by the National land Rights Forum or NLRF (an organization of tenant and landless farmers) and facilitated by a coalition of NGO partners and Community-based Organizations (CBOs). By capitalizing on the support of strategic partners (SPs), the CSRC and its coalition partners channel financial, institutional and technical inputs to the NLRF (including its district and local chapters) embedded with the objective of strengthening and enabling to develop and launch movements/ campaigns from village to the national level. Consequently, a number of achievements have already been made in the past as follows: (i) strengthening the power of land-poor farmers' organization – the village land rights forums (VLRFs); (ii) policy reform ('scientific land reform' has been a major agenda of the state as incorporated in the recently promulgated constitution); (iii) government policy pronouncement for enhancing women's equal access to land (as guaranteed by the new constitution); (iv) community-led land reform; (v) promoting livelihood initiatives of rights through agriculture co-operatives; and (vi) strengthening collaborative actions through alliances and coalitions (Uprety, 2015).

NLRF has been a decade-old institution of land rights holders in Nepal which is an aftermath of incessant CSRC institutional support for its strengthening. In 2015, it had been operational in 53 districts with their district (meso) and village (micro) chapters. Now they have begun functioning independently for land rights policy advocacy and campaigns. Definitely, CSRC has been extending the institutional support (both capacity building and financial). The CSRC has been elected to the board member of the International Land Coalition (ILC) representing South Asia and this event has given it an opportunity to work as the co-chair of the ILC Asia Steering Committee. It has also been a board member of the Asian NGO Coalition for Agrarian Reform and Rural Development (ANGOC). These opportunities have given it the international recognition/space for gaining international support for accelerating the land rights and land reform movement in Nepal. During the partnership period, all international relations maintained during the 2009-2013 have been continued. So is the case with the NLRF. Its ex-chairperson has been the vice-chairperson of the Asian Farmers' Association (AFA) and consequently, the NLRF has now ample space to internationalize the Nepalese issues germane to land rights and agrarian reform in the international arena.

Land rights movement in Bangladesh

Apropos of the land rights movement in Bangladesh, the Association for Land Reform and Development (ALRD), a member of the International Land Coalition (ILC) since 2003, is the federating body of 273 NGOs, peasant and landless organizations in Bangladesh, which are involved in the struggle to establish land rights, rights to food, rights to livelihood, and rights of the indigenous people or minorities. It is currently the main organization in Bangladesh working exclusively on land reform issues. ALRD envisions a Bangladesh where upholding the rights of citizens is the cornerstone of the state and where the state is pro-actively pursuing the promotion and strengthening of the rights of the poor and the marginalized, including the most vulnerable of the society; landless peasants, indigenous peoples, women and religious and other minority communities. ALRD further aspires for a Bangladesh that adopts secularism as a key guiding principle and gender equity and social justice are considered as key objectives of all its undertakings (www.landcoalition. org/en/regions/asia/member/alrd, downloaded on 7-29-2016).

Land rights movement in India

India has a long history of land rights. For example, the birth of the *Bhoodan* Movement (land donation movement) has been associated with Vinobha Bhave, an Indian eminent social activist, in 1951 when he announced the goal of collecting 50 million acres of land for the land-poor. Bhave wanted landowners to redistribute unwanted or uncultivated land. Between 1951 and 1960, Bhave traveled 25,000 miles on foot, persuading 700,000 landowners to give up 8 million acres. This was a substantial achievement even though the movement failed to reach its goal of acquiring one-sixth of India's cultivable land, and much of the donated land was, in fact, wasteland and uncultivable (in Bihar this was nearly half of all the land donated to the movement). Still by 1961, only 872,000 acres of *Bhoodan* lands had been distributed. Later, the *Telengana* movement engaged in armed struggle to claim land from violent and exploitative landowners. The history of modern India is filled with the land struggles of the poor and dispossessed: from the peasant revolts of Avadh during 1919-1922 which resulted in organizing independent *kisan sabhas* (peasants' associations), to the 1967 Naxalbari movement of West Bengal. At its height, the *Telengana* movement succeeded in shutting down the administrative machinery of the Nizam in 4,000 villages, and in establishing *gram rajya*, or village self-rule. The institution of vetti, or compulsory, forced labor was abolished, and 10-12 *lakh* acres of land was redistributed. The *Telengana* movement was also notable for the widespread participation of women, though they did not receive any of the redistributed land in their names unless they were widows (Visweswaran, 2007, pp.5-6).

Of late, the Ekta Parishad (Unity Forum) has been leading the movement for land rights in India for last 25 years. In 2007, with the support of several other groups (like the National Association of People's Movements), it led 25,000 landless *Dalits* and *Adivasis* from 12 different states on a four-week *Padayatra* covering 350 kilometers on foot from Gwalior in Madhya Pradesh to the Indian Parliament in Delhi. Their purpose was to highlight the urgency of land reforms for the poor. The march, called "*Janadesh*," or the "People's Verdict," took three years to organize and had as its objectives: (i) the creation of a National Land Commission with statutory powers to direct state governments to carry out comprehensive land reform; (ii) an end to evictions of *Adivasis* (indigenous) people from forest lands, and (iii) new fast track courts to resolve land disputes quickly. "The Indian state pushed through Special Export Zone (SEZ) legislation in a matter of days, yet in 60 years it had not been able to realize pro-poor land reforms in a country where three-quarters of the population works in the agricultural sector; where farmer suicide was rampant in 100 districts, and Naxal insurgency in another 172 of India's 600 districts", remarked Rajgopal, the leader of Ekta Parisad. The *Janadesh* march participants were displaced *Dalits* and *Adivasis* who had farmed the same land for generations, but had seen it expropriated for large-scale industrial development projects in conjunction with state-sanctioned SEZ schemes. The Ekta Parishad insists that the same principle that underlines its movement, "no conversion of agricultural land for non-agricultural purposes". As a result of the *Janadesh* 2007 mobilization, the Indian government has set up a National Land Reforms Council headed by the Prime Minister with representation of the Ekta Parishad. The *Janadesh* 2007 Satyaghraha built upon long-standing and localized traditions of Gandhian land protest, but has also deliberately sought to internationalize the agenda of landlessness and indigenous land rights by including representatives from Canada as well as international observers from Kenya, Brazil, France, Scotland and several other countries. Up to 50 percent of the *Janadesh* march participants were *Adivasis* evicted from their lands who had cases against them for violations of the Forest Act as they attempted to engage in daily subsistence activities. Activists who work with the indigenous landless in Brazil, having previously organized marches of up to 12,000 people were also interested in the logistics of organizing 25,000 for a march of several hundred kilometers. The *Janadesh* organizers divided march into groups of five thousands; then further into groups of one thousands, and finally into groups of twenty-fives. As part of the self-respect traditions of the movement, each group of twenty-five was responsible for maintaining discipline and for cooking and distributing food among themselves for the duration of the march. The Ekta Parishad, founded in 1991, bases itself on Gandhian principles. One of its slogans has been "Between silence and violence is active non-violence." Its activists have faced police harassment and jail on numerous occasions during the course of their work. One observer has said of the movement's emphasis on *"Rachana, sangharsh aur bahishkar"* (Creation, struggle and boycott) that "its ideological aspiration is to rediscover the radical in Gandhi" erased from the social memory of post-independent India. Ekta Parishad also looks to Jayaprakash Narayan's socialism and takes Vinobha Bhave's *Bhoodan* Movement as one of its models, focusing on the redistribution of illegally occupied bhoodan lands in states like Bihar. As a "new social movement," Ekta Parishad sees itself neither as a trade union, a political party, or NGO, but as a mass-based social movement that works though allied networks across eight different states, many of them states where the Naxal movement is also strong. The organization is the strongest in Madhya Pradesh with 100,000 members spread out over 2,068 villages in 26 districts (ibid, 6-7).

In 2015, the government led by Prime Minister Narenda Modi introduced a Land Act Ordinance by passing from his cabinet. It proposed to exempt five categories of acquisitions from the procedural requirements of the 2013 Act. These five categories were: defense, industrial corridors, rural infrastructure, affordable housing including housing from the poor, and any infrastructure including social infrastructure in public-private partnership mode (PPP) where the land is owned by the government. The implication of government amendment meant that new social impact assessment, food security assessment and consent of 70 percent of landowners would not be required before acquiring land for these five categories of land that the government has specified. Basically, government would fit every acquisition under each of these five categories by annulling the previous Act itself (https://www.reddit.com/r/india/comments/2re3r8/ salient_ features_ of land act ordinance-2015).

In 2015, in solidarity with Anne Hazare, an eminent anti-corruption crusader, Ekta Parishad in collaboration with other social organizations launched a *Yatra* of 5,000 *Adivasis*, farmers and landless from 12 states (a march of 60 kilometers on foot by following the Mathura Road toward the parliament street in Delhi) to oppose the land acquisition ordinance which also got large political support. Finally, the delegation of Ekta Parishad was invited by the Home Minister Mr. Rajnath Singh for putting its proposal. Then, he talked to Mr. Narendra Modi, the Prime Minister, about land distribution to the landless and protection of land rights of marginalized people. The government has agreed to reconstitute the National Land Reform Council chaired by the Prime Minister. In this regard, Rajgopal, Ekta Parishad's leader remarked, "Nearly 50 million people are homeless and cannot live with dignity and possibility to earn livelihood. Land rights were the issues of Ekta Parishad for the past 25 years. I am happy that land issue has been on the political agenda. That is the first issue in the right direction" (www.ektaparishad, 4 March 2015).

Referring to the paper on 'the land grabbing in India', personal correspondence of Nathaniel Marquez, Executive Director of ANGOC, with this author has shared a few critical findings. In India, the village tank land has also been grabbed by the powerful people for the construction of buildings in urban areas and agriculture and brick industries in rural areas. It is not only the *Dalits* who have had the land problems but also the tribal people and Muslims, Christians and the relatively marginalized Hindus. Paradoxically, the lands which the government has acquired in the name of *bhoodhan* movement and land reform measure (which set the land ceilings) until now have not been distributed properly as there is no accountable information regarding the redistribution of land (i.e. who got how much land). It is the politicians who decide who will get the lands. The land which the people got under these initiatives are sold secretly with the cooperation of government officials and migrated to the neighboring towns and cities but no transparent information was available on such malpractices.

Human Rights Institutions and Mechanisms in South Asia

SAARC Charter on Human Rights

Founded in December 1985, SAARC consists of eight countries, namely, Afghanistan, Bangladesh, Bhutan, India, Maldives, Nepal, Pakistan and Sri Lanka which have shared cultural and historical ties for centuries. The objectives of its charter are geared toward promoting the welfare of the people of the region, improving the quality of their life, accelerating economic growth, social progress, and cultural development, providing opportunities to their citizens to lead a life with dignity and to realize their full potential, promoting and strengthening collective self-reliance among the countries, contributing to mutual trust and understanding among them, promoting active collaboration and assistance in the economic, social, cultural, technical and scientific fields, strengthening cooperation among themselves in international forums on matters of common interests and cooperating with international and regional organizations with similar aims and purposes.

Indeed, literature demonstrates that there is initial focus on development initiatives. Since 2002, it took steps to address human rights beginning with the 'convention on preventing and trafficking in women and children for prostitution' and 'convention on regional arrangements for the promotion of child welfare in South Asia'. There are various human rights commitments in SAARC's broad objectives (such as providing opportunities to their citizens to lead a life with dignity and to realize their full potential which have the implications of the citizens' right to health, education, adequate care and adequate standard of living and promoting the welfare of the people of the region including improving the quality of their life). Article 11 of the SAARC charter states, "SAARC shall not be a substitute for bilateral and multilateral cooperation but shall complement them. SAARC cooperation shall not be inconsistent with bilateral and multilateral obligations." All the member countries, except

Bhutan, have ratified the ICESCR, two multilateral treaties at the core of the International Bill of Human Rights along with the Universal Declaration of Human Rights. All the eight countries have agreed to comply with the responsibilities prescribed by the 'Convention on the Elimination of all Forms of Discrimination against Women as well as Convention on the Rights of the Child'. These seven countries have multilateral obligations to reinforce the rights stipulated in these covenants, which are basic human rights. It does not have a human rights body or a treaty for cooperation of its members on issues related to the International Covenant on Civil and Political Rights (ICCPR) and ICESCR (Sattar, Seng and Muzart, 2012, pp.24-25).

In 2004, SAARC's social charter was signed with focus on 'poverty eradication, population stabilization, empowerment of women, youth mobilization, human resource development, promotion of health, and protection of children'. Its preamble states, "The principal goal of SAARC is to promote the welfare of the peoples of South Asia, to improve the quality of their life, to accelerate economic growth, social progress and cultural development, and to provide all individuals the opportunity to live in dignity and to realize their full potential". SAARC's social charter can be interpreted along the wide range of economic, social and cultural rights. One can discern the broad commitment to upholding human rights in South Asia. One of its objectives is to "promote universal respect for the observance and protection of human rights and fundamental freedoms for all, in particular, the right to development; promote the effective exercise of rights and the discharge of responsibilities in a balanced manner at all levels of society; promote gender equality; promote the welfare of children and youth and promote social integration". In the context of economic, social and cultural rights, Article 3.4 states, "State Parties agree that access to basic education, adequate housing, safe access to drinking water, and sanitation, and primary health care should be guaranteed in legislation, executive and administrative provision, in addition to ensuring of adequate standard of living, including adequate shelter, food and clothing". Article 3.5 states the imperative of providing a better habitat to the people of South Asia as part of addressing the problems of the homeless. Indeed, the charter is the potential foundational tool for regional human rights initiatives. There is also mention of 'food security', and the establishment of 'food bank' in 2007 for tackling the food shortages through the 'regional food security reserve' and provisioning of 'regional support to national food security efforts' and fostering 'inter-country partnership' to tackle regional food shortages through the collective effort (SAARC Social Charter, 2004 and Sattar, Seng and Muzart, 2012, pp. 36-37).

SAARC and Land Rights

As indicated above, SAARC policies have underscored the issues of poverty alleviation, improvement of agricultural production and attainment of food security. In 2008, ANGOC took the initiative to assess the land issues in South Asia in its 'Land Watch Asia'. A concise analysis is presented underneath on the land issues of the region through the examination of its findings.

Given the fact that poverty has been put at the centre and pro-poor strategies have been adopted as per the recommendation of the Independent South Asian Commission on Poverty Alleviation, SAARC's development goal on livelihood had also recommended the distribution of land to the landless in the region. Indeed, the ambitious goal was to halve the number of poor people until 2010. Two targets were set to realize this goal: (i) the proportion of population below the calorie based food plus non-food poverty line, and (ii) distribution of state land to landless tenants. Indeed, the possibility of realizing the target of land distribution has been mere 'lip service' in the absence of any reliable benchmarks as indicated earlier on. The SAARC group on livelihood had underscored that targeting would require macro-economic and sectoral approaches for poor people's sustainable livelihood which does include natural capital (such as land and water). On the one hand, there is ambiguity on SAARC's position on the importance of land rights and the absence of official declaration from SAARC on land rights and issues pertaining to the farmers in the region, and on the other, SAARC's social charter ignores the land issues (despite the fact that the majority of the region's citizens are rural poor farmers). Nonetheless, SAARC appears to recognize the significance of land distribution in the context of large number of landless people in the region for meeting the SAARC's development goal on livelihood. There has been the realization among the member countries about the continuing decline in the availability of land, a critical resource for agricultural development. Indeed, the mention of the provision on food security in SAARC's charter can be linked to land issues (given the fact food security and nutritional security can be possible with the availability of land). ANGOC's paper shows that the dominance of growth-oriented framework fails to clarify the following issues: (i) does SAARC put a premium on land rights and issues raised by farmers when it says that it aims to "improve the quality of life in South Asia"? and (ii) does SAARC believe that South Asia can proceed with tackling other development projects without first resolving land issues? It is also not clear on its plan of action for poverty alleviation whether SAARC regards land as one of the resources to which the poor have no access. In other words, there is no elaboration of land rights/issues in its program on poverty alleviation. There appears to be a lack of awareness of land as a basic problem. This could embody a number of questions such as: (i) whether access to land and land rights per se are considered main issues by SAARC?; (ii) how does SAARC define or perceives land issues?; (iii) what priority is given by the SAARC officials to land issues? and (iv) whether they view land rights as an inter-related or separate issue, *inter alia*? There is no mechanism for CSO participation in such discussions on the primacy of agricultural development and the need to ensure food security (ANGOC, 2008, pp.1-4). The realization of equitable economic development is contingent only when access of landless/land-poor to land and land tenurial security is ensured (as the organization has set the goal of distributing land as a poverty alleviation target which has the potential of addressing food security also but it suffers from the mere 'lip service').

Some Initiatives of SAARC Country Members: Special Rapporteurs on Agrarian Issues and National Inquiries on Indigenous People's Rights

Literature shows that to date there has been no collective initiatives of SAARC country members through the appointment of Special Rapporteurs on agrarian issues and national inquiries on rights of indigenous peoples who constitute a sizable population in South Asia. For instance, indigenous peoples comprise 37 percent in Nepal, 15 percent in Pakistan, 8.6 percent in India, and 1-2 percent in Bangladesh. For these peoples as elsewhere, land is as culture and survival who have been disenfranchised by the national expansionist and colonial governments in the past. Then, the state-sponsored assimilation and state-led migration also had negative bearing on the customary practices of land use among the IPs (i.e. disappearance of such practices over time). Gradually, the national governments framed discriminatory state policies and promoted practices to implement them. Of late, there has been 'new colonialism' protected by the state policies for the extractive industries and plantations and national development activities (Quizon, 2015) which have negative bearing on the land and natural resources of IPs.

The review of SAARC documents also shows that the Technical Committee on Agricultural and Rural Development (TCARD) does not have the clear stance of the organization on the panoply of land-related issues. Therefore, with the appointment of the Special Rapporteur for agrarian issues and the national inquires on the rights of IPs vis-à-vis land issues, a concerted institutional effort may be made in the future to focus on: land rights, agrarian reform (redistributive policies), program of access to productive resources such as land, tenurial rights, sustainable use and management of common property resources (such as forests, water, genetic resources, biodiversity and land), resettlement and relocation, access to legal instruments for land disputes, women's rights to land, customary rights of IPs, stakeholder participation in formulating agrarian reform policies, ILO 169 agreement, agrarian reform in places of conflict/war, etc. (ANGOG, 2008).

In the context of South Asia, UN human rights instruments such as ICESCR provide the foundation for the recognition of customary land of the IPs. And under such condition, there is the need to undertake the common agenda and action for their rights, providing and fulfilling the regional level UN Declaration on the Rights of IPs through the initiation of dialogues among CSOs in the region to address the land and agrarian issues through concerted policy advocacy, learning and exchanging on policy development (through sharing experiences and best practices), learning from specific country experiences, working for holistic reforms on land and resource governance and recognizing the IPs as key to our collective future for conserving eco-systemic resources, maintaining biodiversity and promoting indigenous knowledge systems (Quizon, 2015).

UN Guiding Principles on Business and Human Rights

The Human Rights Office of the High Commissioner of the United Nations prepared and published the document entitled 'Guiding Principles on Business and Human Rights' in 2011 with a view to implementing the United Nations position to "protect, respect and remedy" the violations or abuses of human rights in the member countries. It has clearly spelled out the state's duty to protect human rights and apropos of it, there are two foundational principles: (i) states must protect against human rights abuse within their territory and/or jurisdiction by third parties, including business enterprises (which requires taking appropriate steps to prevent, investigate, punish and redress such abuse through effective policies, legislations and adjudications), and (ii) states should set out clearly the expectation that all business enterprises domiciled in their territory and jurisdiction respect human rights throughout their operations. It has also spelled out the operational principles vis-à-vis general state regulatory and policy functions. While meeting the duty to protect human rights, states should: (i) enforce laws that are aimed at, or have effect of, requiring business enterprises to respect human rights and periodically assess the adequacy of such laws and address any gaps; (ii) ensure that laws and policies governing the creation and ongoing operation of business enterprises including corporate laws do not constrain but enable business respect for human rights; (iii) provide effective guidance to business enterprises on how to respect human rights; and (iv) encourage, and where appropriate require, business enterprises to communicate how they address their human rights impacts (Human Rights Office of the High Commissioner, 2011, pp.4-5).

Referring to the state-business nexus, the document also admonishes the states to take additional steps to protect against human rights abuse by the business enterprises, and suggests them to exercise adequate oversight in order to meet the international human rights obligations and promote respect for human rights. It also underscores that states and businesses must not be involved in heightening the gross abuses of human rights in the conflict-affected areas.Under such grim situation, they must be engaged for the identification, prevention and mitigation of the abuse or violation of human rights. It has also suggested the states' access to public support and services for business enterprises that are responsible for gross human rights violations (pp. 9-10).

The document also emphasizes the role of states (which have to act as members of multilateral institutions) for ensuring policy coherence among their departments, agencies, and other state-based agencies for the protection of human rights. It has also articulated the foundational and operational principles for the corporate responsibility to respect human rights. Under the foundational principles, business enterprises must avoid infringing on the human rights of others, fulfill the responsibilities to respect internationally recognized human rights (including rights of rights of indigenous people and women), seek to prevent or mitigate adverse human rights that are directly linked to the services/operations of business enterprises, etc. These enterprises should also have policy commitment to meet their responsibility to respect human rights as underscored in operational principles. Besides, they have also laid stress on human rights due diligence to identify, prevent, mitigate and account for how they address their adverse human rights impacts, and issues of transparency, accountability and remediation through legitimate processes (pp.13-23). Apropos of issues of context, business enterprises are asked for compliance with laws and respect for internationally recognized human rights, honoring the principles of internationally recognized human rights, treating the risk of causing gross human rights abuses, and prioritizing actions for addressing actual and potential adverse human rights impacts. Regarding access to remedy, the foundational principle asserts that all human rights abuses must have access to effective remedy (through the redress of grievance – a perceived injustice evoking a sense of entitlement). And its operational principles underscore the importance of state-based or domestic judicial mechanisms, state-based non-judicial grievance mechanisms (administrative, legislative, and appropriate non-judicial grievance mechanisms alongside judicial mechanisms), and non-state based grievance mechanisms (adjudicative, dialogue-based or other culturally appropriate and rights-compatible processes) for having access to remedy. Finally, the effectiveness criteria for non-judicial grievance mechanisms (both state-based and non-state based) should be legitimate, accessible, predictable, equitable, rights-compatible, and a source of continuous learning. Operational level mechanisms should be based on engagement and dialogue (pp. 31-34).

Indeed, these foundational and operational principles are also equally important in the regime of land rights. It also underscores the necessity to have greater clarity in the laws and policies governing the access to land, subsuming entitlements germane to ownership or use of land, with a view to protecting rights-holders and business enterprises. Definitely, there can be issues of gender, vulnerability and marginalization of indigenous, minority, and scheduled castes in the context of South Asia tied with the access and ownership of land.

Bringing Land Rights in HR Mechanisms in South Asia

In discussing the issues to bring land rights in human rights (HR) mechanisms in South Asia, it is essential to shed light on the condition of CSOs and their human rights mechanisms in South Asia. CSOs have proliferated in the last 30 years and have attained vibrancy in their activities. Although there exists an institutional culture of collaboration between the governments and CSOs on a panoply of development issues, government authorities and human rights activists are at loggerheads because the latter are found to be raising the issues of human rights violation and atrocities committed by the government authorities against the people clamoring for the protection of rights (be they political, civil, social, cultural, and economic).Human rights activists/defenders have been the targets of harassment, intimidation, arbitrary detention, torture, and even extra-judicial killings. Thus, the SAARC countries have a relatively weak record of human rights promotion. However, CSOs have been seamlessly found to be involved in their activities for defending human rights. They have also the SAARC as a regional platform to make their voices heard (Sattar, Seng and Muzart, 2012).

The SAARC social charter also has regard for civil society because it reaffirms the need to develop, beyond national plan of action, a regional dimension of co-operation in the social sector. It also espouses principles that members of the civil society uphold, such as equity and social justice; respect for and protection of fundamental rights; respect for diverse cultures and people-centered development. But the official documents are silent on the accreditation of CSOs (SAARC Social Charter, 2004 and ANGOC, 2008).

Since the early 1990s, a number of human rights organizations have come into existence. For instance, the South Asians for Human Rights (SAHR) is a membership-based regional organization working for the protection of human rights, peace-building and democratic progress. There is another organization called South Asian Forum for Human Rights (SAFHR) which works as a forum for dialogue between regional and local human rights organizations. In 1994, a new regional entity came into being called 'People's SAARC' – a collective movement of South Asian civil societies since 1994. It discusses ways to foster cooperation at people-to-people level in South Asia when the official SAARC process fails to address the issues (SAARC Social Charter, 2004 and ANGOC,2008). The process of People's SAARC has firmly established and a tacit consensus on its significance and collective ownership built among the South Asian activists. Hence, whatever groups or organizations have the spirit and resources to pool into its organizing and logistics, civil society and common people welcome it. The changing name of the event – People's SAARC, South Asian People's Summit, People's Assembly – is an indication of its organic, spontaneous and inclusive nature (PSAARC India, 2013). The first people's SAARC meeting held as parallel event to the 8th SAARC summit in New Delhi in 1995 to lobby SAARC on the issue of trafficking and then the 9th SAARC recognized trafficking as a grave issue in 2002 (in the form of convention). Since then, People's SAARC has been functioning to lobby SAARC officials on regional concerns. The 10th SAARC resolved to establish the South Asia Forum to serve as a platform for debate and exchange of ideas at a regional level between government representatives and stakeholders. In 2010, an organized institutional effort was made to establish a Working Group on South Asia Human Rights Mechanism in Kathmandu as an outcome of the regional gathering of human rights activists sponsored and organized by Forum Asia and the Informal Sector Service Centre (INSEC). It produced the Kathmandu declaration calling for the establishment of "an independent, effective and accountable human rights mechanism with explicit mandates of promoting, protecting, and fulfilling human rights through a process of wide consultation with NGOs, people's movement at national and regional levels'. In 2011, a working group was established. On the whole, SAARC seems to be in the incipient stage of the enforcement of human rights (Sattar, Seng and Muzart, 2012). Nonetheless, land rights issues can be brought to the People's SAARC, South Asia Forum and the Working Group on South Asia Human Rights Mechanism for debates and discussions and its findings can be communicated to the SAARC governments through the SAARC secretariat.

Linking Land Rights to SDGs

It is increasingly recognized that human rights are essential to achieve sustainable development. The Millennium Development Goals (MDGs) had served as a proxy for certain economic and social rights but ignored other important human rights linkages. By contrast, human rights principles and standards are now strongly reflected in an ambitious new global development framework, the 2030 Agenda for Sustainable Development. On 25 September 2015, 170 world leaders gathered at the UN Sustainable Development Summit in New York to adopt the 2030 Agenda. The new Agenda covers a broad set of 17 Sustainable Development Goals (SDGs) and 167 targets and will serve as the overall framework to guide global and national development action for the next 15 years. Grounded in international human rights law, the Agenda offers critical opportunities to further advance the realization of human rights for all people everywhere, without discrimination. SDGs are different in their nature. Unlike MDGs which were applied to 'developing countries' only, SDGs are truly a universal framework and has been applicable to all countries. All countries have progress to make in the path towards sustainable development, and face both common and unique challenges to achieving the many dimensions of sustainable development captured in the SDGs. They are transformative because as an agenda for *"people, planet, prosperity, peace and partnership"*, the 2030 Agenda offers a paradigm shift from the traditional model of development. It provides a transformative vision for people and planet-centered, human rights-based, and gender-sensitive sustainable development that goes far beyond the narrow vision of the MDGs. They are comprehensive because alongside a wide range of social, economic and environmental objectives, the 2030 Agenda promises *"more peaceful, just and inclusive societies which are free from fear and violence"* with attention to democratic governance, rule of law, access to justice and personal security (in Goal 16), as well as an enabling international environment (in Goal 17 and throughout the framework). It, therefore, covers issues related to all human rights, including economic, civil, cultural, political, social rights and the right to development. They are also inclusive because the new Agenda strives to leave no-one behind, envisaging *"a world of universal respect for equality and non-discrimination"* between and within countries, including gender equality, by reaffirming the responsibilities of all states to *"respect, protect and promote human rights, without distinction of any kind as to race, colour, sex, language, religion, political or other opinions, national and* social origin, property, birth, disability or other status" (www.ochr.org/ En/Issues/MDG/Pages/The 2030 Agenda.aspx, downloaded on 11-8-2016:1). The Human Rights Office of the High Commissioner of the United Nations had also made an important contribution to integrate human rights in the entire process of defining SDGs for ensuring that strategies and policies to implement the 2030 Agenda are human rights-based.

More specifically, the **preamble** of the resolution on SDGs has recognized that **eradicating poverty** in all its forms and dimensions, including **extreme poverty**, is the greatest global challenge and an indispensable requirement for sustainable development. There has also been a resolve to free the human race from the **tyranny of poverty and want** and to heal and secure the planet. There is determination to take the bold and transformative steps which are urgently needed to shift the world on to a sustainable and resilient path. There is a pledge that no one will be left behind in the collective journey. The emphasis of SDGs has been the quest to realize the human rights of all and achieving gender equality and the empowerment of all women and girls. They are integrated and indivisible and balance the three dimensions of sustainable development: the economic, social and environmental. The Goals and targets will stimulate action over the next 15 years in areas of critical importance for humanity and the planet (UN General Assembly, 2015).

Below is an overview of the goals and targets of SDGs as specified in the resolution on September 25, 2015 by the UN General Assembly which have their relevance to land rights as human rights as analyzed in the preceding sections. More specifically, **Goal 1** of SDGs is "**no poverty**: end poverty in all its forms everywhere" and the target is to ensure that all men and women, in particular the poor and the vulnerable, have equal rights to economic resources, as well as access to basic services, ownership and control over land and other forms of property, inheritance, natural resources, appropriate new technology and financial services (including microfinance) by 2030 (UN General Assembly, 2015).

Goal 2 is "zero hunger: end hunger, achieve food security and improved nutrition and promote sustainable agriculture" and the targets are: (i) to double the agricultural productivity and incomes of small-scale food producers, in particular women, IPs, family farmers, pastoralists and fishers, including through secure and equal access to land, other productive resources and inputs, knowledge, financial services, markets and opportunities for value addition and non-farm employment, and (ii) ensure sustainable food production systems and implement resilient agricultural practices that increase productivity and production, that help maintain ecosystems, that strengthen capacity for adaptation to climate change, extreme weather, drought, flooding and other disasters and that progressively improve land and soil quality by 2030 (UN General Assembly, 2015).

Goal 5 is 'gender equality: achieve gender equality and empower all women and girls' and the target is to undertake reforms to give women equal rights to economic resources, as well as access to ownership and control over land and other forms of property, financial services, inheritance and natural resources, in accordance with national laws by 2030 (UN General Assembly, 2015).

Goal 15 is 'life on land: protect, restore and promote sustainable use of terrestrial ecosystems, sustainably manage forests, combat desertification and halt and reverse land degradation and halt bio-diversity loss". The main targets of the above goal are: (i) to ensure the conservation, restoration and sustainable use of terrestrial and inland freshwater ecosystems and their services, in particular forests, wetlands, mountains and drylands, in line with obligations under international agreements by 2030; (ii) promote the implementation of sustainable management of all types of forests, halt deforestation, restore degraded forests and substantially increase afforestation and reforestation globally by 2020; (iii) combat desertification, restore degraded land and soil, including land affected by desertification, drought and floods, and strive to achieve a land degradation-neutral world; (iv) ensure the conservation of mountain ecosystems, including their biodiversity, in order to enhance their capacity to provide benefits that are essential for sustainable development; (v) take urgent and significant action to reduce the degradation of natural habitats, halt the loss of biodiversity and, by 2020, protect and prevent the extinction of threatened species; (vi) promote fair and equitable sharing of the benefits arising from the utilization of genetic resources and promote appropriate access to such resources, as internationally agreed; (vii) take urgent action to end poaching and trafficking of protected species of flora and fauna and address both demand and supply of illegal wildlife products; (viii) introduce measures to prevent the introduction and significantly reduce the impact of invasive alien species on land and water ecosystems and control or eradicate the priority species by 2020; (ix) integrate ecosystem and biodiversity values into national and local planning, development processes, poverty reduction strategies and accounts for: (a) mobilizing and significantly increasing financial resources from all sources to conserve and sustainably use biodiversity and ecosystems; (b) mobilizing significant resources from all sources and at all levels to finance sustainable forest management and provide adequate incentives to developing countries to advance such management, and (c) enhancing global support for efforts to combat poaching and trafficking of protected species, including by increasing the capacity of local communities to pursue sustainable livelihood opportunities (UN General Assembly, 2015).

The above depiction of four major SDGs shows that their realization by 2030 would be possible provided landless people, marginalized women and IPs have access and control to land and other economic resources. Complete eradication of poverty and hunger and ensuring gender equality would only be possible once the land rights of land-poor people are addressed by each member state of UN and SAARC.

Conclusions

Based on the analysis presented in the preceding sections, three broad conclusions have been drawn:

(i) From the rights-based perspective, civil, political, economic, social and cultural rights of human beings are recognized as universal, inherent, inalienable, indivisible and interdependent body of rights as empirically experienced. Politically, people have to have the right to self-determination which allows them to make their own independent decisions for the free pursuit of their economic, social and cultural development. There can be no deprivation of the means of subsistence for the people under any circumstance. Institutionally, the promotion of the realization of the right to self-determination is the responsibility of the state and therefore, there can be progressive realization of the rights through the adoption of legislative measures for their enjoyment without discrimination. Undertaking efforts to ensure the equality among men and women to enjoy all economic, social and cultural rights must universally be the priority of the states. Recognition of the rights by state parties to citizens' adequate standard of living (subsuming adequate food, clothing, housing and to the continuous improvement of living conditions), fundamental right of freedom from hunger and cultural rights is of paramount importance. Thus, there is the indivisibility of human rights and land rights can be considered as an 'accessory right' to the realization of other human rights. Violation of land rights of people leads to the violation of human rights (be they civil and political or social, economic and cultural). For instance, in Nepal, this would include the right to avail of government services (e.g. water, electricity, banking services, citizenship certificates which are vital for the enjoyment of voting rights and also serve as the key to acquiring passports for gainful foreign employment which constitutes 25 percent of gross national income in the form of remittance).

(ii) The realization of equitable economic development is contingent on the access of landless/land-poor to land and ensuring of land tenurial security (as the SAARC organization has set the goal of distributing land as a poverty alleviation target which has the potential of addressing the food security also but it suffers from mere 'lip service'). The inextricability and interconnectedness among the land rights, poverty alleviation, agricultural development and food security cannot be denied at any cost. The ambivalence of SAARC's policy on land issues, lack of clear regional strategy for enhancing the land-poor people's access to land and other productive natural resources (in societies characterized by skewedness in landholding), culture of patriarchy and colonial and national government exploitative/discriminatory policies towards the exploitation of resources (owned by IPs) have triggered the perpetuation of resource inequity in the SAARC region.

(iii) Given the fact that land issue has not been collectively addressed by the member states of SAARC through the formulation of a common regional strategy, "social justice on land" has been a far-fetched dream. They now have state obligations, both in the capacity of members of 'comity of nations' and individual independent states, to respect and protect the rights of land-poor farmers (landless and marginalized including women and IPs living in poverty) on land (including homestead land) and fulfill such obligations by changing the state policies/laws for ensuring the "equitable land distribution" and implementing them responsibly (including preventing and remedying land-grabbing) as specified under 'people-centered land governance' of the International Land Coalition (ILC) for translating the goals of social justice into realities in foreseeable future.

Recommendations

Based on the analysis furnished above, the following recommendations have been made:

Bringing the land rights issues in the People's SAARC (South Asian People's Summit, People's Assembly), South Asia Forum and Working Group on South Asia Human Rights Mechanism (the only existing mechanisms) for debates and discussions and communicating the findings to the SAARC governments through the SAARC secretariat would be appropriate.

Creation of a platform on which the government representatives of all member countries of SAARC can participate in the dialogues of land rights issues raised by prominent CSOs and representatives of victims of the gross violation of land rights in the region to convince the SAARC authorities to create permanent inter-governmental human rights mechanisms such as the Regional Committee on the Issues of Land Rights and Special Rapporteur on Agrarian Issues and Land Rights of Indigenous Peoples for minimizing the gross violation of land rights in the region and ensuring the enjoyment of civic, political, economic, social and cultural rights of the people would be equally important (SAARC as a provider of enabling environment for the well-being of the land-poor people such as landless, marginalized women farmers and IPs).

The regional network of CSOs should play a proactive role in engaging the Technical Committee on Agriculture and Rural Development (TCARD) created in 2006 (to address the challenges for ensuring the food and nutritional security at the level of agricultural ministers in the SAARC) on land-related issues.

Collaboration between universities and prominent national human rights organizations working on land rights in each country for the conduct of empirical researches for national evidence-based robust advocacy is of paramount importance. The formation of a regional entity for collaboration between these institutions to give feedback to the SAARC secretariat/governments is also highly recommended.

Contributions have to be made by governments, CSOs and academic institutions of the SAARC region for enabling land-poor people to ensure their land rights (as human rights) by linking their national and regional programs on land resources for the accomplishment of SDGs such as "No poverty" (goal 1), "Zero hunger", (goal 2), "Achieving gender equality" (goal 5), and "Life on land" (goal 15).

Enhancement of collaborative efforts of national governments, CSOs, CBOs and research organizations in the SAARC region for "achieving gender equity" for land rights through women awareness creation and empowerment programs, holding policy dialogues with key stakeholders (responsible for decision-making) for formulating pro-women land polices/laws, conducting gender-sensitive researches vis-à-vis women's land rights and ensuring the policy implementation through lobbying and monitoring is also highly recommended.

Note: The paper was originally written for Asian NGO Coalition for Agrarian Reform and Rural Development (ANGOC) in 2016 with its financial support grant in a slightly different theme entitled "Mainstreaming Land Rights as Human Rights: An Approach Paper". Now the paper has been revised by using the lens of "engaged anthropology", a sub-field of applied anthropology. The author has taken the permission from Mr. Nathaniel Marquez, the Executive Director (ED) of ANGOC for the publication of the original paper in a slightly different way. The author extends sincere thanks to ANGOC and its ED for the sponsorship and permission for publication. Mr. Jagat Basnet, Ph.D scholar in Sociology from Rhoades University of South Africa, deserves sincere appreciations for his first perusal of the draft and furnishing the suggestions.

References

Agrawal, B. (1996). *A field of one's own: Gender and land rights in South Asia.* New York: Cambridge University Press.

Agrawal, B. (1999). Gender, property and land rights: Bridging a critical gap in economic analysis and policy. *Out of the Margins,* 264-289.

ANGOC (2008). Land issues in South Asia: Assessing SAARC. *Land Watch Asia,* Issue Brief 1. Manila: The Philippines.

ANGOC (2014).2014 *Land reform monitoring report: Towards an accountable governance on land in Asia.* Manila: The Philippines.

_____. (2015). Women land rights in Asia: *Land Watch Asia.* Issue Brief. Manila: The Philippines.

Barkat, A. (2015). *Land reform report 2014: Bangladesh.* Association for Land Reforms and Rural Development (ALRD) and Human Development Research Centre (HDRC), Unpublished.

Barkat, A. et al. (2015). *Assessing inheritance laws and their impact on rural women in*

Bangladesh. Conducted for International Land Coalition.

Bachriadi, D. (2014). Land rights as seen from human rights perspective: A challenge for the Asian community. *Workshop on the theme 'mainstreaming land rights as human rights' (16- 17 September, 2014)*. Phnom Penh, Cambodia.

Bhardwaj, S. (2014). India situation. *Workshop on the theme 'mainstreaming land rights as human rights'* (16-17 September, 2014), Phnom Penh, Cambodia.

Bourdreaux, K. and Sacks, D. (2009). Land tenure security and agricultural productivity. *Mercatus on policy* No. 57. George Mason University.

Coalition on World Food Security (CFS) and UN Food and Agriculture Organization (FAO) (2012): *Voluntary guidelines on the responsible governance of tenure of land, fisheries and forests in the context of national food security*. Rome, Italy.

Central Bureau of Statistics (CBS), 2011. *Nepal living standards survey*. Vol.1. Kathmandu: Nepal.

_____. (2011). *National living standards survey.*Vol.11: Kathmandu.

_____. (2012). *Poverty in Nepal (Brief report based on NLSS 111-2010/2011)*. Kathmandu.

_____. (2012). *National population and housing census 2011* (National Report). Kathmandu.

Charya, C. (2014). Welcome. *Workshop on the theme 'mainstreaming land rights as human rights'* (16-17 September, 2014). Phnom Penh, Cambodia.

Chaudhary, S. n.d. *Land grabbing in India for food security*. A Microsoft Power Point Presentation, Housing and Land Rights Network and Ekta Parishad, India.

Clarke, C.M. (2010). Toward a critically engaged ethnographic practice. *Current Anthropology.*Vol. 51. Supplement 2.Pp. 301-312.

Constituent Assembly of Nepal (2015). *Constitution of the federal republic of Nepal*. Kathmandu.

Corpuz, V. T.2015. *UN special rapporteur on the rights of indigenous peoples on the impact of international investment and free trade on indigenous peoples' rights*. Full Report Presented to 70th Session of UN General Assembly on 7 August, 2015.

Daily Star, May (5, 2015). www.thedailystar.net/package/rural-women-deprived-land-ownership-80534.downloaded on 7/17/2016.

Dhar,V.(2012). *Agricultural census in India*. Delhi: Department of Agriculture and Cooperatives, Ministry of Agriculture, Government of India.

Diliman declaration of the Asian people's land rights tribunal forum, 16-17 January, 2014, Malcom Hall Theater, University of the Philippines, Diliman, Quezon City, The Philippines.

Gefont, (2014). *Transformation: By building workers' power.* 6th National Congress, March 28-31, 2014, Kathmandu.

Ghimire, K. (1998). *Forest or farm: The politics of poverty and land hunger in Nepal.* New Delhi:Manohar.

Gonzalez, N. (2010). Advocacy anthropology and education working through the binaries. *Current Anthropology.*Vol. 51. Suuplement 2.pp.259-267.

https://landwatchasia.wordpress.com/tag/land-rights-movement-in-India,downloaded on7/19/2016.

https://www.reddit.com/r/india/comments/2re3r8/salient_ features_ of land act ordinance-2015, downloaded on 7/19/2016.

His Majesty's Government of Nepal, (2005). *National agriculture policy.* Kathmandu.

Human Rights Office of the High Commissioner, *Report.19 and 20 info notes (2014) based on the summary of the report of the special rapporteur Leilani Farha on the right to adequate housing.*

Humanrights.americanathro.org/1999-statement-on-human rights.

IFAD, (2012). *Land tenure reduction and poverty reduction.* Retrieved from the IFAD Website: http://www.ifad.org./pub/factsheet/land.e.pdf.

Jeremie, n.d. *Land rights are human rights: The case for a specific right to land.* University of East London.

Johnston, B.R. (2010). Social responsibility and the anthropological citizen. *Current Anthropology.*Vol. 51. Suuplement 2.Pp.235-247.

Kardam, N. (1998). Changing institutions in women's interests. In *Development and Gender in Brief 5.*

Kellett, P. (2009). "Advocacy in anthropology: Active engagement or passive scholarship" *Durham Anthropology Journal,* Vol. 16 (1) 22-31.

Landesa. (2015). *Status of women's land rights in three countries of South Asia (India, Nepal, and Pakistan): A draft consolidated briefing paper.*

Landesa. (2015). *Status of women's land rights in Nepal: A draft briefing paper.*

Integrated Food Policy Research Institute (2013). *Status of food security in the feed the future zone and other regions: Results from 2011-12 Bangladesh integrated household survey.* Report Submitted to USAID, Dhaka, Bangladesh.

Islam, S; Moula, S; & Islam, M.(2015). Land rights and land disputes and land administration in Bangladesh: A critical study. *Beijing Law Review,* 6, 193-198.

Low, S.M. and Merry, S.E. 2010. Engaged anthropology: Diversity and dilemma: An introduction to supplement 2. *Current Anthropology.*Vol. 51. Suuplement 2. Pp.203-225.

Open-ended Intergovernmental Working Group of Human Rights Council on *'the rights of peasants and other people working in rural areas'*, 17-20 May 2016.

Oxfam (2014). *Women farmers: Challenges and opportunities.* Lalitpur, Nepal.

Oxfam (2014). *Annapurna Gatha: Struggle launched by women of different districts for food and identity and leadership in agriculture sector and achievements in their own words.* Lalitpur, Nepal.

Oxfam (2014). *Proceedings of women farmers' national assembly.* Lalitpur. Oxfam.

Phnom Penh Communiqué on Land Rights and Human Rights in ASEAN *at the conclusion of the workshop on the theme 'mainstreaming land rights as human rights"* (16-17 September 2014).

Program on Covenant on Economic, Social and Cultural Rights (PWESCR), (2015): *International covenant on economic, social and cultural rights:A handbook.* Jan.P.Strijboslauan: The Netherlands Office, The Netherlands.

PSAARC India, (2013). *Evolution and history of people's SAARC.* India (Downloaded on 13 July 2016).

Quizon, A. B. (2013). *Land governance in Asia: Understanding debates on land tenure rights and land reform in Asian context.* Rome: ILC.

Quizon, A. B. (2015) On the customary land rights of indigenous peoples in Asia. *Land Watch Asia:* Issue Brief.

Rawal, V. (2014). Changes in the distribution of operational landholdings in rural India: A study of national sample survey data. *Review of Agrarian Studies.* Vol.3, No. 2, July, 2013- January 2014.

Regmi, M.C (1971). *Land ownership in Nepal.* Delhi: Adroit Publishers.

Report of the Special Rapporteur on *'Adequate housing as a component of right to an adequate standard of living'* by Miloon Kothari, 25 January 2001.

Report of the Special Rapporteur on *'Adequate housing as a component of right to an adequate standard of living'* by Miloon Kothari, 1 March 2002.

Report of the Special Rapporteur on *'Adequate housing as a component of right to an adequate standard of living'* by Miloon Kothari, 3 March 2003.

Report of the Special Rapporteur on *'Adequate housing as a component of right to an adequate standard of living'* by Miloon Kothari, 8 March 2004.

Report of the Special Rapporteur on *'Adequate housing as a component of right to an adequate standard of living'* by Miloon Kothari, 8 March 2005.

Report of the Special Rapporteur on *'Adequate housing as a component of right to an adequate standard of living'* by Miloon Kothari, 14 March 2006.

Report of the Special Rapporteur on *'Adequate housing as a component of right to an adequate standard of living'* by Miloon Kothari, 7 February 2007).

Report of the Special Rapporteur on '*Adequate housing as a component of right to an adequate standard of living and on the right to non-discrimination in this context'* by Miloon Kothari, 13 February 2008.

Report of the Special Rapporteur on '*Adequate housing as a component of right to an adequate standard of living and on the right to non-discrimination in this context'* by Raquel Rolnik, 4 February 2009).

Reports of the Special Rapporteur on '*Adequate housing as a component of right to an adequate standard of living and on the right to non-discrimination in this context'* by Raquel Rolnik, December., 2010 and 26 December 2011.

Report of the Special Rapporteur on '*Adequate housing as a component of right to an adequate standard of living and on the right to non-discrimination in this context'* by Raquel Rolnik, 24 December 2012.

Report of the Special Rapporteur on '*Adequate housing as a component of right to an adequate standard of living and on the right to non-discrimination in this context'* by Leilani Farha, 22 December 2014.

Regmi, M.C. (1999). *A study in Nepali economic history: 1768-1846.* Delhi: Second Reprint, Adroit Publishers.

SAARC, (2004). *SAARC social charter.* Islamabad, Pakistan.

Sattar, M. Seng, Y.S. and Muzart,S. (2012). *SAARC and human rights: Looking back and ways forward.* Bangkok: Asian Forum for Human Rights and Development (Forum Asia), Thailand.

Sourav, R.I. (2015).Unjust land rights of women in Bangladesh *International Research Journal of Interdisciplinary and Multidisciplinary Studies* .Vol.1, 5-13.

Susser, I. (2010). The anthropologist as social critic: Working toward a more engaged anthropology. *Current Anthropology.*Vol. 51. Suuplement 2. Pp.227-233.

United Nations Human Rights Office of the High Commissioner (2011).*Guiding principles on business and human rights: Implementing the United Nations "protect, respect and remedy",* Geneva, Switzerland.

United Nations General Assembly (1948).*United Nations universal declaration of human rights.* Adopted and Proclaimed by General Assembly 217 (111) of December 1948.

United Nations General Assembly (1966). *International covenant on economic, social and cultural rights.* Ratification and Accession by the General Assembly Resolution 2200A (XXI) of 16 December, 1966; Entry Into Force 3 January 1976.

United Nations General Assembly, 2015. *Transforming our world: The 1930 agenda for sustainable development.* New York: USA.

Uprety, L. P. (2014). *Ensuring equitable access to land through land rights movement.* Kathmandu: Community Self-Reliance Center.

_____. (2015). *State of women farmers in Nepal: A conspectus:* A Final Report Submitted to Oxfam Country Office, Kathmandu, Nepal.

_____. (2016). *A critical ethnography: Empowerment of land-poor farmers through the movement for land rights and agrarian reform in Nepal.* Kathmandu: Community Self-Reliance Center.

Velayudhan, M. (2009). Women land rights in South Asia: Struggles and diverse contexts. *Economic and Political Weekly,* Oct. 31. XLIV (44).

Virak,Ou. (2014). Welcome. *Workshop on the theme 'mainstreaming land rights as human rights" (16-17 September, 2014).* Phnom Penh, Cambodia.

Wickeri, E and Kalhan, A. (2010). Land rights issues in international human rights law. *Malasiyan Journal of Human Rights,* Vol. 4, No.10.

Wiswesworm, K. (2007). Contemporary land rights movement in India. *Occasional Papers of the Sabaltern Popular Workshop.*1-7.

www.ektaparishad, 4 March, 2015 downloaded on 7/20/2016.

www.visionias.in.n.d. *Socio-economic and caste census 2011 in India.* Downloaded on 7/18/2016.

www.ohchr.org/EN/Issues/Housing/Pages/Housing Index.aspx. downloaded on 6/5/2016.

www.org/EN/Issues/Food/Pages/FoodIndex/aspx downloaded on 7/27/2016.

www.ochr.org/En/Issues/MDG/Pages/The 2030 agenda.aspx. *Human rights and the 2030 agenda for sustainable development.* Downloaded on 11-8-2016.

Zaman, M.A. (1973). *Evaluation of land reform in Nepal.* Kathmandu: Ministry of Land Reform.